INNOVATION
AT
SCALE

Strategies and Actions of Visionary and Innovative Companies

ANKITA VASHISTHA
AVINASH VASHISTHA
(Bestselling author of THE OFFSHORE NATION)

ALSO BY AVINASH VASHISTHA
The Offshore Nation

First published by Tholons Inc. in 2021.

e-book ISBN: 978-1-952067-04-4
(USA $9.99 / CAN $12.99 / UK £6.99)

Paperback ISBN: 978-1-952067-05-1
(USA $19.99 / CAN $25.99 / UK £14.99)

Hardcover ISBN: 978-1-952067-02-0
(USA $29.99 / CAN $38.99 / UK / £21.99)

Tholons Inc.

For volume purchases and discount please contact
book@tholons.com or +1 646 875 4945 or +1 408 454 8611

To Our Innovation Conscious Families,
especially Garima, Sanath, Abhay, Amisha and Aveer

Contents

Preface

In the past decade, the digital revolution has taken industries by storm—and no sector is immune to disruption. New digital technologies have given birth to a new business age—the "Age of Innovation." This book covers how digital is impacting industries across financial services (banking), healthcare/pharmaceuticals, retail/consumer goods, education, manufacturing, media/entertainment and energy. The most prominent technologies that are disrupting the industries are robotics, artificial intelligence, cognitive computing, IoT and cyber security. We look at how, the business operations of these companies are being impacted significantly by digital.

Enterprises will realize "Innovation at Scale", through "Framework for Innovation at Scale" model, that brings in re-imagining customer experience, transformation through innovative solutions and intelligent automation. The book also highlights some of the major innovations by global leaders, and how women empowerment, engagement and entrepreneurship is accelerating digital transformation and will add USD 12 Trillion by 2025 (McKinsey). This book profiles the top 100 Super Cities and 50 Digital Nations that are leading the charge globally.

The book is intended for both C-level executives and practitioners—professionals who are making and implementing the digital transformation decision. However, digital transformation is a megatrend that has the potential to affect everyone's life, both personally and professionally. It has the potential to enrich all our lives. Therefore, we hope that academics, economists, politicians, consultants, investment bankers, manufacturers, industrialists, services sectors and students all recognize that these topics are directly applicable to their future.

For all these reasons, we bring you *Innovation at Scale.*

Ankita Vashistha
Avinash Vashistha

Author's Bio

Avinash (Avi) Vashistha is Chairman and CEO of Tholons and Managing Director at MyStepUp Entrepreneurship and Innovation Foundation and StrongHer Capital. He is the former Chairman and CEO of Accenture (India). Avi is a highly accomplished, results oriented Chief Executive (CEO) and Venture Capitalist (VC) with 30+ years of experience in large global projects, CEO/board level strategy consulting, venture capital, digital transformation, and innovation.

As Chairman of Accenture, he led a multi-billion-dollar business and a workforce of over 160,000 delivering a >3X growth in 5 years. Prior to that he was the founder of Tholons and neoIT and earlier with Nortel and AT&T in US and UK. Avi has done over 100 large globalization deals valued at over USD 50 billion. He has led investments in over 30 companies and has been a Limited Partner in 5 Venture Capital/Private Equity Funds.

Avi has an exceptional network and relationships with top 1000 Chairman/CEOs, Bureaucrats, Ministers and policy makers across the USA, Europe-UK, Asia-India, and Latin America. He is a known Thought Leader in the industry and co-authored the book "The Offshore Nation" (McGraw-Hill). He has worked with over 500 Global Fortune clients like Goldman Sachs, GM, Chevron, Pfizer, Kaiser Permanente, Vodafone, T-Mobile, SAP, Stanford University, TATA, Bank of America, UBS, SBI, Timken, GE, Walmart, Shell, Exxon Mobil, BMW, Unilever, The Home Depot, Accenture, VISA, CardinalHealth, JP Morgan Chase, P&G, PayPal, Pepsico, John Deere and others.

Ankita Vashistha is the CEO of Tholons Capital, MyStepUp Entrepreneurship and Innovation Foundation and Managing Partner at StrongHer Capital, with over 12 years of experience in private equity, consulting, venture capital and innovation across UK, US and Asia. She is an engineering graduate and masters from Cranfield school of management, UK. Ankita works very actively in the startup ecosystem globally across US, UK, India, Singapore and Japan, to source, evaluate, mentor and invest in tech companies. She also works with portfolio companies to help them create value, scale and expand, leveraging technology.

Ankita founded and launched India's first venture capital fund, Saha Fund, to promote and invest in women entrepreneurship and technology. The investment portfolio is across e-commerce, healthcare, education, tech platforms, food tech, mobile, data analytics, enterprise solutions, artificial intelligence, cloud etc. Some of her notable investments across her funds are Instahealth, Luxola, Licious, J2W, Fitternity.

Previously, Ankita had worked at Aureos Capital. At Aureos Capital, Ankita was part of the global investor relations and fund portfolio team. She helped raise funds for Africa Healthcare fund, India Fund II and Southeast Asia Fund II totaling to more than USD 500 million. Ankita was also part of the founding team for Wavemaker Labs, Singapore. Ankita is also part of many angel networks, academic mentor programs and is part of the NASSCOM product council.

Acknowledgements

Ankita Vashistha

Education and global experience have made me the free and independent person I am today. I always say that I never want to stop learning, because the day I do, I will stop growing as a person. It's so important to travel and meet people and expand our horizons. Without either, we will never be able to tap into the infinite opportunities out there. I have been extremely fortunate to have lived across the world since a child, having lived and studied across US, UK and Asia. This, has not only truly made me an independent person but also sensitive to different cultures and people. It allowed me to see the way different worlds work and how we can all make an impact with our experience and education.

My diverse experiences have allowed me to think how I can really impact those around me. Having lived in Asia, and being witness to the widening income inequality, the contrast between urban and rural, makes you think of frugal innovation. Makes you think of what an immense impact new technology can make in the field of education and healthcare. How real technology can address the demand and needs of a billion people and provide quality services and products from anywhere in the world. Implementation and reach of 5G and the extensive mobile network, where even the low-income labor worker has a smart phone with an internet connection. Makes you really think of what best practices you can bring to a region and other emerging markets.

How can we empower women in a world, where still only a single digit percent of women represents startups and companies at leadership positions? How can we get 50% of the population actively involved in the working ecosystem? How can we bring in all the amazing innate qualities a woman has, and inculcates in the multiple roles she takes on during her lifetime, to the workplace? This thought pushed me to think about diversity and inclusivity and I launched India's first venture capital fund to invest and promote in women entrepreneurship, employment and engagement. To create gender balance in business, we need to create a huge wave of women entrepreneurship and get everyone involved in thinking about it. We must make an impact by not just thinking, but by actually doing and implementing.

All my thoughts and inspirations are a culmination of meeting and being able to work with the most amazing people all around. I would love to take this opportunity to thank all the special people in my life. My teachers at Mallya Aditi International School, Bangalore, India who made sure I grew up to be a free and independent thinker. Mr. Ram Bhat and Mrs. Banerjee. My most amazing mentors at Aureos Capital, who gave me my first private equity job—Brigit Van Dijk—who inspired me every day to come to work with her never say never spirit, Noah Kim—his eye for detail made me perfect at details. The amazing women leaders who inspired me to launch a women entrepreneurship fund and were my early believers—Kiran Mazumdar Shaw and Zia Mody. My colleague at Tholons—Anthony Rajesh, who has been an amazing leader and support to all our initiatives and driving this book. Special thanks to Dr. Garima Sathi Vashistha for her contribution of "Heathtech – Redesigning the future of staying healthy" chapter in this book.

Last but not the least, my beautiful family. My parents, who have always believed in me and supported me to embrace my strengths and weaknesses. My mother Garima's strength and determination has always pushed me to move forward. My father has been a true inspiration at home and at the workplace, making sure he steps up every day for and helps everyone around him to grow. He has set inspirational goals for us to achieve. My brother Abhay, the avid footballer and sportsman and my sister Amisha, the multitalented social media queen, who are like my children, my lifelines and who are my go-to buddies for everything. My extremely supportive and passionate husband Sanath, who has supported me and inspired me every day to take on more risk and achieve more and my little bundle of joy, Aveer, for who I want to touch the stars and bring the moon and make everything colorful, like a rainbow.

This book will give you an insight into emerging technologies and industries and the innovation happening across the world. This book will give you a sneak peek into the world of technology, entrepreneurship and innovation and create curiosity within you to discover and lead change.

Avi Vashistha

"Leadership" exits because of people who have the vision, the conviction, the passion, the entrepreneurship and the fearless selfish zeal to do the "right" thing. "Stewardship" is sweet—to mentor future leaders and entrepreneurs, making the world a better place for the next generation! Entrepreneurship and Innovation are the lifeline of the digital economy.

Steve Jobs has long been an inspiration for me personally—an entrepreneur, exemplary innovator, a leader and an amazing steward! He will always live in my heart and soul as my drive, passion and zeal for life.

I am very thankful to my former colleagues at Accenture and my vast network of friends at leading consulting, technology, Fortune 500 and Global 2000 companies. They have been my closest partners to engage, challenge and enjoy the excitement of technology transformation journeys over the course of last two decades—more so over the last five years on digital.

This book would not have been possible without the idea, drive and co-authoring of this book by Ankita Vashistha Shetty, who is an exemplary leader having formed the first Women Entrepreneur fund in India (Saha Fund), after an illustrious career in UK and US with Aureos Capital and Tholons Capital. She has bought in her experience of working with Innovation platforms globally and with the startups in India, Singapore, Japan, UK and USA.

Special thanks to my trusted colleagues at Tholons, particularly—Anthony Rajesh, who has been my partner in transformational initiatives, globalization strategies and the thought leadership. I am grateful to Frank Pendle, who has been my partner in the Intelligent Automation journey, for his contribution towards "Intelligent Automation" chapter in this book. Avnish Sabharwal was my partner in Accenture in establishing the open Innovation group and he led that initiative to great heights globally. I am thankful to Avnish for authoring an amazing chapter on Open Innovation. Great insights and guide from Venkat Thiruvengadam for his contribution on the chapter "Cloud and Infrastructure As A Service."

My pillar of strength and inspiration has always been my lovely family—my dad, Ram Kishor Vashistha—a visionary leader, my mom Urmila—on how to raise four kids, each of whom are at the helm of their profession. My lovely wife, Garima (m. 1984), who is wiser than I am, a greater risk taker and an excellent decision maker and always a willing partner to go along with my adventures. She has been my biggest support for every remarkable achievement of my life including this book!

I do learn and shape my thinking and my path based on what I learn from my three lovely children—Ankita (the global innovation platform expert and a Venture Capitalist), Abhay (Avid Sportsman, Sports Entrepreneur and a Sports Management Professional) and my youngest, Amisha (doing her degree at LSE in Management/Digital Innovation, but a Chef, Fashion aficionado, Marketing marvel at heart).

I have to recognize few of my mentors that are very close to my heart and I owe my success to them: Kevin Campbell, who has been the reason for my success with my first two entrepreneurial ventures (neoIT and Tholons) and my most rewarding career at Accenture. Diju Raha, who was the founder of "Outsourcing" and Globalization of Services for giving me the pioneering opportunity to be involved at the foundation and startup of the Global Outsourcing phenomenon and Industry. Amelia Rowen, at Nortel for mentoring me to be an exemplary leader and giving the opportunity of the international assignment. I have really been fortunate to get such love, guidance and opportunity from so many of you! A true blessing!

Introduction

In the past 25 years, the digital revolution has taken industries by storm, and no sector is immune to disruption. This technological hurricane is forcing economies to be reshaped overnight as digital disruption brings down flourishing businesses to ground zero in no time.

Following along the lines of the initial analogy between the unpredictability of nature and technology, it should be pointed out that human have a knack for creating order out of chaos. New digital technologies have given birth to a new business age—the "Age of Innovation." The advent of new technologies like social media, analytics, big data, mobility, IoT, cloud, intelligent automation, robotics, cognitive computing, as-a-service, artificial intelligence and virtual reality are only some, overturning incumbents and reshaping markets faster than ever. Technology is redefining industries and enabling enterprises to diversify their product and service offerings. The only means to survive this competitive business ecosystem is through "Framework for Innovation at Scale" model, that brings in re-imagining customer experience, transformation through innovative solutions and intelligent automation.

The genuine challenges, therefore, lie not in combating disruption, but in turning these inevitable digital disturbances into opportunities—pinpointing the ways in which technology is revolutionizing industries, and identifying how organizations, leaders and economics can react to it. Enterprises need to learn to develop their digital strategy—protect themselves by embracing digital innovation at the agility of a startup. You definitely, cannot beat them and need to join them. Governments across the globe are aligning policies, regulations, tax incentive systems and other initiatives to boost entrepreneurship and the open innovation ecosystem.

The book 'Innovation at Scale', co-authored by Ankita Vashistha and Avinash Vashistha offers an empowering approach to instilling the awareness of how industries, countries and lives of citizens have been reshaped forever. The work comprises a thorough guide on how to successfully navigate the sea of abundant new digital technologies across various industry sectors and outlines a strategic roadmap for enterprises and countries to evolve as leaders in this world of new digital technologies and collaborative innovation at scale.

1
Digital Disruption

"We lived on farms, then we lived in cities, and now we're going to live on the INTERNET!"
— **Sean Parker**

In this day and age, the internet ranks exceedingly high on the list of things we take for granted in our lives. What is now a whole universe in itself, comprised of terabits of data and information, had a very humble start. It was by no means intended to conduct any e-commerce transactions. It was launched by scientists in Europe in order to share ideas and concepts with fellow scientists around the world.

In order to get into how the internet and other technologies changed the way we live; we need to understand how they actually work. What happens when a customer clicks on the "Buy Now" option on Amazon? How does an Uber driver know the location of the person booking the cab? And how do dating websites know about my perfect match better than my mother? The answer is a simple word taught in every 4th grade computer class—"algorithms."

Algorithms exist everywhere. If we define economics as a study of people's behavior in a given environment, then algorithms are the processes that go into such behavior. They are sets of rules that control everything from genetic code to computer code. We cannot live without them. For instance, back in the days, we had to step out of the house and go find a cab to use the taxi service. The first innovation was the taxicab callboxes, and then came the two-way radio dispatching which gave way to computer-assisted dispatching, and now we have companies like Uber. We can have a cab come to us by the touch of a button. The person booking the cab does not even have to tell the driver what their destination is, the Uber App does it for them. The client just enters the destination from their phone, and it syncs with the driver's phone. The app uses GPS to navigate the streets and get the client to their destination via the most time-efficient route.

The best part about Uber is that it is crowd sourced. This means that Uber does not own the cabs. People drive their own cars and get paid for it. Uber saw the cost and hassle of direct transportation as a problem that it was, and decided to solve it. Most successful businesses in the world work this way. They identify problems and then provide solutions for it. This process of identifying and fulfilling previously unmet needs is called value creation.

Another algorithmic innovation is shaping our daily lives. India's Aadhaar is the world's largest biometric ID system, with over 1.2 billion enrolled members. They are currently using biometric data to serve the public. People can access useful services such as mobile phone and cooking gas connections, as well as banking by having an Aadhaar number. While in its inception the number was considered only for government-run services, the near future will see other similar services by non-governmental entities come into its fold. The Aadhaar project has already been linked to many public subsidies and schemes that benefits citizens. Aadhaar cards are now being linked to bank accounts and biometric machines at ATMs for hassle-free payments. In theory, they are aiming to reduce the red tape to a number. As we can see, even this model is solving a problem and, hence creating value.

However, not all businesses are created out of the need to solve an issue. Facebook is a great example. Initially, Facebook was not solving any problems. The public was not hungry for a new method of communication, and there were other websites such as MySpace already doing almost exactly what Facebook was about to do. So, what did Facebook do differently that made it a leader in social media networking?

It made its user profiles exclusive. Users would have to know a person to be granted access to their profile. This is what we today know as Facebook's "Friend requests." Studies have shown that humans tend to migrate towards things they cannot have. Facebook's exclusivity factor allowed the exclusivity algorithm to work. People would have to know each other to accept "Friend Requests," thus perpetuating the atmosphere of privacy and exclusiveness.

Unwanted advertisements accompany the browsing of free websites. They disrupt the general flow of user experience and ultimately the very objective of marketing. But the ads, of course, are there for a reason—they are the company's source of revenue. Besides, thanks to digital tracking technology, ads these days are tailor-made and targeted. Marketers no longer have to rely on assumptions about consumer behavior but can instead target individuals with ads based on their actual behavior online. Naturally, this type of personalized advertising is more effective at generating both clicks and conversions. Everyone with an internet connection has been served an ad, prepared for them especially, no matter what they were using the internet for. Google is the largest online ads revenue generator in the world, and websites like Google and Facebook have the largest exposure in terms of customers visiting their site, which makes them prime places for advertising. Facebook had no ads in the beginning. Mark Zuckerberg did not want ads to take virtual space that was intended for communication and connection. So how did they get so popular? Not so much by solving a problem as by finding a niche in people's mentality. In the words of Zuckerberg—*"People can go anywhere on the internet and see pictures of girls. They came to our website because they wanted to see the pictures of girls they know. Why not offer a site that does exactly that?"* And so, Facebook was born.

Zuckerberg's basic idea was that people wanted to go online and look at what their friends were doing when they were not with each other. This curiosity he sensed in the general population for what is now widely recognized and more pessimistically called "the fear of missing out" has enabled the rise of the greatest social network in the history of mankind. Zuckerberg disrupted the forever old concept of stepping out of home and talking to other people by using nothing more than his computer. Oh… and his brain, of course.

Many companies have managed to disrupt the established flow of business by playing with digital frontiers. Amazon is one of them. Amazon started off when Jeff Bezos, the founder of Amazon decided to combine the timeless and physical with the new and digital. He took orders of books online and then went to deliver the books to their respective buyers. This small business model has now grown to be the largest sellers of merchandise online. People are attracted to such sites for multiple reasons. The ease of transactions and time saving are at the top of the list. People can browse through catalogues on their computer and are just a click away from purchasing the product. These websites also offer products for cheaper prices as there is a significant reduction in middlemen involvement. The product goes from the seller's warehouse to the buyer's doorstep in a matter of days. In fact, Amazon today offers same day delivery on many items, even on weekends. While e-commerce does have its drawbacks like every other business model. Some customers simply prefer to feel their prospective purchases and engage in human face to face interaction. However, these websites are giving brick and mortar stores—a run for their money and heavily disrupting how consumers shop.

It is not just the websites that have excelled in the use of technology to disrupt daily lives. There are other products which have been digitally enhanced to perform better and make people's lives easier. One such example is the Roomba, the automatic vacuum cleaner. Not only does it clean for you, it is self-operated. All it needs is to be switched on and it cleans the floor on its own. It uses a spatial algorithm integrated with motion detectors to detect nearby obstacles, adjust its course and vacuums areas where there is no furniture or other interferences.

Moving off the floor into the air, the highly talked-about drones are taking various markets by storm. One that has fallen prey to commercial drone invasions particularly hard is the photography market. Drones allow photographers to reach new heights with their camera skills. Shots can be taken at previously unattainable angles and elevations with little effort. Drones are even used to explore nature's hidden secrets. Researchers and photographers exploring ancient caves and large holes in the earth use them to explore the area first; making sure it is safe to venture further.

Amazon has taken the drone concept a step further and started deploying drones to deliver packages to restricted areas. This project is still under test, and it may take a while for it to become part of everyday life. Similarly, Google has been testing cars that could drive themselves, which seems like one of its biggest diversification strategies. Digital technologies are enabling enterprises to diversify their products and services, drive higher revenues and expand their consumer base. Most consumers are unaware how busy these companies are. Facebook is testing Terragraph which augments terrestrial cellular networks with millimetre-wave technology that delivers data 10 times faster than existing Wi-Fi network technology. An online food ordering and delivery platform was founded in August 2014 called Uber Eats, which is a subsidiary of Uber. Microsoft's initiative "Moon Shoot" has researchers analyzing our genetic code, finding ways to reprogram the immune system to combat cancer cells more effectively. Along with it, "Bio Model Analyzer," a software tool used to figure out why leukaemia patients respond differently to different treatments, is used by AstraZeneca and Microsoft.

Zooming out to the global level and going back to the advent of the digital age, even as early as the 70s and 80s, conducting businesses across time zones was a hassle, especially between developed countries like the United States and developing countries like China and Japan. The time difference alone made it difficult for businessmen to conduct meetings, not to mention the travel it took to meet budding clients. A lot of money and time was spent on uncertain outcomes. Companies like Cisco and Skype succeeded in changing the way humans communicate for good as people from different parts of the world can talk to each other via video, voice call or chat for free. In addition, multiple people can log into multiple devices and use the service at the same time, developing effective, synergistic approaches to problems without having to allocate resources.

Businessmen can also get real-time updates on the state of markets through various applications. Certain applications are designed to give its users live updates on stock markets and market conditions so that brokers can make quick decisions to improve their chances of success.

Business intelligence is increasingly becoming a necessity in all workplaces. Big data is the backbone of decision making. Going digital allows easy calculations and presentation of such data. Companies like Marketo and Salesforce allow its users to make smarter decisions in the marketplace by giving them an edge in decision-making processes. Even the hard-core process of marketing and lead generation is made easier through the use of these platforms. Both Marketo and Salesforce have a direct email feature which allows companies to send bulk emails to a large customer base with the press of a button. It also analyses the emails and provides information on whether they provoked the desired responses.

And so, in the last ten years, technology has disrupted the way people behave in their everyday lives. With the advent of artificial intelligence, users no longer have to go through pages of data to find what they are looking for. The device does it all for the user. Some popular examples of artificial intelligence are "Siri", "Alexa" and "OK Google." Users can now talk to their computers and give them instructions. Artificial Intelligence is designed to replicate human behaviour, making it easier for humans to interact with their devices, and, in turn, make choices that define their relationship with their environment.

Technology has certainly affected the way people run business today. A self-funded startup turned Tech Company promoting accessible education called Study.com beautifully defines technological change as *"The improvement in the art of making products or developing processes."* It often takes a lot of creativity to run a business; a technological change alleviates this by making efficiency increasingly more achievable. Businesses around the world are thus slowly, but surely migrating towards the capital side of the factors of production and moving away from the human resource side. Both the Large companies and the small businesses can use technology as a part of their core competence. By using computers, servers, websites, personal digital products and sometimes crowdsourcing (products and services available to almost everyone) to develop an advantage in the economic environment through their own competence.

For instance, Airbnb, one of the largest hospitality companies, does not own a single property. It crowdsources its resources, thus cutting down its own costs immensely. The company has been disrupting the hospitality industry ever since it was launched. It is a service operating through a website that allows people to rent their own rooms, apartments and houses. People seeking temporary accommodation can visit this online marketplace, type in the desired vacation/temporary stay location, and filter out the types of houses they do not want to stay in. The rents at such place ranges from incredibly low to pretty high, but are always cheaper than traditional accommodation, and offer hidden gems and authenticity one could only experience while staying with friends/relatives.

Big hotel chains around the world like the Marriott's and the Hyatt's have lost a significant amount of business to Airbnb as people tend to opt for the cheaper and more customizable options. Airbnb, established in 2008 is now valued at over USD 38 billion, overtaking the market value of prominent hotel chains. Airbnb has also established an online reputation system where guests and hosts can rate their experience for future reference. This allows for a trustworthy relationship among the users. Uber, whose drivers operate as hosts of their own cars, exchange feedback with their clients in the same way. Crowdsourcing leads to some level of community in an open market.

The use of technology and its increasing ease of user experience have overflowed into some of the more rigid business environments. The financial sector is known to be one of the most rigid business sectors, with stock brokers and bankers using age old methods to conduct business—until recently. The arrival of fintech, which is basically the application of technology in financial processes, the financial industry has revolutionized for good. Clients neither must go all the way to the bank to check their balance nor do they have to use cash and chequebooks to conduct transactions.

Businesses and banks around the world are moving towards a cash-free economy, where transactions can be conducted with ease and in record time. A simple illustration of this is PayPal, Google Pay, Ali Pay, Apple Pay and Paytm. These systems work in a way that there is no cash deposit anywhere at all—everything is virtual. When a user pays through their Google Pay account, the funds get transferred directly from the banking institution to the recipient. There are many wallet-based accounts, where users add money to their wallet by using credit or debit cards. There is no liquid cash that exchanges hands.

The medical industry is another field where technology has stepped in and disrupted the flow of regular business. The merger of technology and human medicine have saved countless lives around the world. Biotechnology and information technology have played a significant role in improving the health and healthcare of people around the world. In the field of medical research, technology has been allowing scientists to go ever deeper. They have been examining diseases on a cellular level and producing antibodies to fight them with extensive use of technology.

Physicians, patients and healthcare providers are witnessing benefits of new medical technologies. The use of electronic medical records (EMR), telehealth services and even mobile technologies like tablets and smartphones is on the rise in the industry. The integration of medical equipment technology and telehealth that defines the distribution of health-related services and information electronically has made robotic surgeries possible. In some cases, it eliminates the need for a physician to be in the operating room with a patient when the surgery is being performed.

As new technologies get implemented in hospitals and research centres, the laws of using it need to be repeatedly updated. Regulations like HIPAA and its Privacy and Security Act addresses the concerns about the confidentiality of patient information. Medical providers must make sure that any new technology and its services are "HIPAA approved" before investing in their implementation.

Even the retail market has fallen prey to innovative digital disruption. The shift from traditional shopping behaviour and the rising popularity of online retailers have already been mentioned, but there is much more to it. There has been a significant rise in show rooming, the use of social media for product reviews, instant feedback and access to price comparisons, as well as the integration of smartphones with other retail technologies. In Steven Keith Platt's words, *"Retailers are striving to target messaging and marketing to their customers to create a unique, personalized shopping experience."*

Retailers are also attempting to collect a massive amount of data on shopping experiences to come up with effective strategies to increase sales. A major challenge to this approach is harnessing big data and carrying out an analysis that makes the data meaningful. Once the data is harnessed and turned into a source of usable information, retailers can move forward with a strategy to build the stores of the future. The traditional way of shopping has long given way to "on the go tech-savvy shoppers" way of research where they browse, purchase, pay wherever and whenever they please. That, in turn, demands the retailers to be more sophisticated in a way they predict demand, manage and move inventory, as well as integrate their physical, virtual and mobile selling channels.

Among those that are impacted the most by technology was the energy industry. Energy efficiency is a burning subject, and while outdated technologies fall to measure up to cost-efficiency and ecological standards, new technologies are not quite ready to power everything from smartphones to cars. Consequently, intermediary technologies have been developed to smooth over the transition in this field of utmost importance to the human race.

We all know what coal is. It is pencil tip's and diamond's carbon cousin. The reason why most energy companies use coal, despite it being a hazard to the environment, is because it is cheap, easy to find and humans have mastered mining it. When carbon is burned, it releases carbon dioxide which makes the earth warmer and causes pollution. The silver lining to this black-cloud could be Future Gen 2.0. It is neither an app nor a website, even though it sounds like one. It is a near zero emissions power plant being built in Illinois. More than 90% of its carbon emissions will be captured and stored.

Another example of revolutionary technology successfully implemented in the energy industry is fuel-free cars. Electric cars used to only be driven by kids, remotely. Some humans dream big, however. Tesla, for example, is a company that has come up with a new range of electric cars. The challenge of moving from an internal combustion engine to no engine at all—consists batteries not producing enough torque and energy to power a full-sized vehicle. The game changed when researchers developed the lithium-ion battery that store massive amount of energy to operate a vehicle.

Harnessing energy using fossil fuels has been the norm in the energy industry for generations. Due to the harmful effect this practice produced on the environment, companies are turning to alternative methods. One such way to generate electricity is to harness the motions of the ocean. A power plant in Maine, USA, has installed turbines in the ocean that turn with the movement of waves and tides. The first turbine will operate on its own for a year, generating 150 kilowatts for the grid as water runs through the turbine at about 11.3 kilometres per hour. This project is now generating 500 kilowatts from the turbine generator unit and is said to annually generate 2.6 to 3.5GWh.[1]

A very popular use of technological innovation to venture forward and develop a competitive advantage takes the shape in engaging with social media. Social media sites are not only ideal platforms for building and promoting a brand but also the perfect sources of big data and user statistics. Companies like Facebook, Twitter, Instagram and LinkedIn have become business package essentials. According to Fortune Magazine, *"Social media is the ideal way to keep your company successful. Not incorporating Twitter, Facebook and other social media channels into your strategy in this era is the equivalent of insisting the web was just a fad a decade or so ago."* For the younger generation, there is nothing special about using social media. They grew up with it, and it is as easy as riding a bike. An increasing number of them are making a significant profit out of their own personal brands they have gradually built online. However, for most of the CEOs and decision makers, social media remains a confusing factor. They cannot seem to fathom the far-reaching ways in which social media can contribute to their business. The presence of customers on social media is a great opportunity for businesses to know their customers and create brand awareness among potential customers. Remember, Facebook and the fear of missing out? People are on social media because they want to know about each other and tell each other their stories. Besides, the competitors will be present and sharing their story too, which provides great opportunities for businesses to develop an informed customers' strategy and develop their business plan more accurately.

Businesses can also look into Artificial Intelligence (AI) for their business needs. AI is commonly implemented as behind-the-scenes algorithms, able to process big data in order to accomplish a range of relatively trivial tasks, far more efficiently than humans. Startups from all over the world have been developing AI for various industries. For instance, Kensho is a startup that claims to be "The world's first computational knowledge engine for the financial industry." A system uses massive parallel statistical computing, natural language inputs, big data and machine learning to answer complex financial questions posed in plain English.

Even the medical industry is putting AI to use. IBM's MSK trained Watson for Oncology system. For instance, one can parse patient information written in plain English and draw on multiple big data sources to deliver ranked treatment recommendations with links to supporting evidence. AI has been seriously challenging the way humans conduct business by potentially replacing humans in the business world. However, now, these kinds of systems seem more likely to synergize with, rather than replace humans.

The cloud is another tool that companies are looking towards to make their lives easier through augmenting human processes in business. The cloud is a pool of computing resources like servers, storage, applications and voice services that is provided as needed to businesses from a provider's network, eliminating the need for on-site equipment, maintenance and management.

Enterprises are adopting cloud computing for many reasons, primarily business performance resourcing, business agility, rapid go-to-market and cost reduction. The Open Group emphasizes that often, *"There is no single reason why a company might choose to use cloud computing. The decision depends on a complex combination of reasons, rather than being based on a single factor."* Opting for cloud computing usually delivers greater agility, reduced costs and a better quality of service.

The use of cloud computing will change a company's risk posture and profile. Using the public cloud, for instance, can let businesses avoid a large investment in IT resources. Even with risks associated with cloud computing, the cloud is being widely used by businesses.

Since the day man first made tools and weapons, there has been a constant struggle for innovation. If we can find a better way to perform a task, we tend to go for it. This not only makes our life easier, but it also disrupts the way we are used to living. Yes, such innovation of processes and technology disrupts the way businesses are conducted, but the benefits far outweigh the drawbacks in any given situation. Some of us like the disruption more than others. Some initiate it, some have trouble adapting it. However, if there is a better way to perform the task, we should reprogram ourselves— and grow. Dan Millman wisely said that the *"Secret of change is not to focus your energy on fighting the old, but on building the new."* If we embrace change, we will have more energy to eliminate the negatives and amplify the positives.

"Digital is the main reason just over half of the companies on the Fortune 500 have disappeared since the year 2000"–Pierre Nanterme, Former Chairman and CEO of Accenture.

The above quote by the CEO of Accenture should send chills down the back of every businessman who is not willing to make digital innovations a part of their businesses. It is by no means necessary to change your business core competency and strategy to digital, but it is necessary to make digital a part of your overall business strategy.

Digital forces are shaping businesses. Cost factor is now addressed using intelligent automation, and cloud technology. To benefit from these trends, companies need to upskill their labour pool to work in collaboration with intelligent machines.

What often happens is that organizations start to address the digital revolution by digitizing their channels in the front office, i.e. marketing, sales and services. This, however, is not a sustainable strategy as the communication with the back office is inevitably a must. This results in the company's failure to generate profitable business. It is critical to address both the front and the back office to achieve growth through a digitalized business model across the entire organization. Let us break it down further.

What is a digital business? A digital business delivers growth and results by creating unique customer experiences through new combinations of information, business resources and digital technologies that produce innovative outcomes designed to deliver exemplary client and consumer experiences.

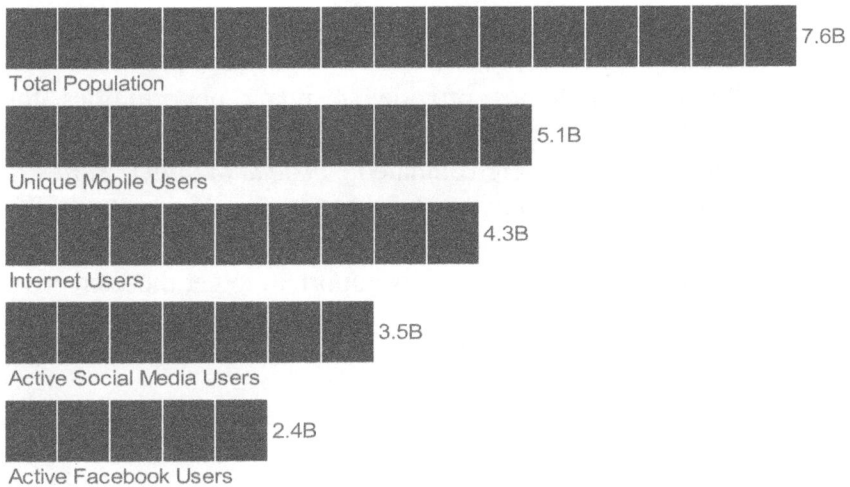

Total Population — 7.6B
Unique Mobile Users — 5.1B
Internet Users — 4.3B
Active Social Media Users — 3.5B
Active Facebook Users — 2.4B

Figure 1: Global Digitization

These numbers have made industries recognize the World Wide Web as a platform worth considering. Global digital advertising spend continues to swell with predictions from Juniper Research estimating that USD 520 billion will be spent by 2023 on top of the current USD 294 billion spend.[2] Social media websites such as Facebook are utilizing their audience knowledge to offer advertisers highly accurate targeting.

Digital transformations are making new impossible things possible every day, so keeping up with the changes is imperative in order to stay on top of things.

For instance, a small marketing company in Bozeman, Montana, United States is revolutionizing the way brands market themselves on digital platforms. They do this by using three simple steps that build around a company's core competency—Research, Build and Refine. In the research phase, the company empathizes with their target audience to understand who they are. After all the target customer is not interested in what they are selling, then there is no need for the product. After obtaining the data, it is compared to the company's web analytics, competitive benchmarks, SEO/SEM data and secondary industry information. Once all the data is collected, a digital prototype is built to test ideas with real users. Once the users have been exposed to the result of the Build phase, feedback is gathered, and the entire process is refined to achieve maximum compatibility.

Digital technologies provide so many opportunities for generating innovation that even the concept of thinking out of the box is starting to be considered outdated. *"Instead of thinking outside the box, get rid of the box,"*—Deepak Chopra. If you have the resources to pull off your ideas, you should worry about the box later. In the beginning, however, it is important to get those creative ideas flowing. Set them free of the box once you have gotten to know your ideas and the box well, both.

Customer-centric businesses are focusing on the customer experience factor of the project. Business requires constant growth, especially in this day and age when technology affects the business and customer side, as well as their interaction. Tell your story, ask for theirs. And learn to speak the digital language. Let's look at some of the services and tools that are shaping and mapping customer experience in today's digital world.

2

Re-imagining Customer Experience

"You have got to start with the customer experience and then work your way back towards the technology, not the other way around."
— **Steve Jobs**

Customer experience (CX), is a journey starting from the discovery of a product/service. Historically, customer satisfaction has been the most critical element of measurement. Today, the focus is on the customer's experience from the conceptualization of a service, to its design and the experience of consuming the same. The fast pace of innovation is changing the expectations and what the consumers get to experience.

Customers respond to direct and indirect contact with a company in different ways. Direct contact usually occurs when the purchase or usage is initiated by the customer. Indirect contact implies blogs, ads, news reports, unplanned encounters with sales representative's word-of-mouth advice and criticism. Brands such as LinkedIn, Amazon, Facebook, and Uber have been incorporating the CX aspect of business into their daily operations. This is illustrated by the absolute number of returning customers. At the end of the day, the product, a business is offering tends to remain the same. It is the customer experience that can be re-imagined and re-designed to make the product more attractive. Technology is enabling businesses to enhance customer experience in various fields. Let us dig a bit deeper into some of them.

Shopping

Today's shoppers have an array of options; shopping in a store, shopping online, getting product reviews, friend's opinion and opting for product comparisons. These benefits are all just a click away. To create a shopping experience that entices shoppers to come back again and again, enterprises need customer experience analytics. They give retailers unprecedented insight into shopping experience from the consumers' perspective.

A majority of companies rank "Improved customer satisfaction" as the primary reason for adopting analytics. Analytics gather data and help the companies understand the needs and expectations of their customers better. This data endorses informed decision making which helps management take manageable risks and make amends.

Large retail stores like Macy's are using big data to offer more localized, personalized and smarter retail customer experience across all channels. Datafloq is a website that analyzes different data points; such as out-of-stock rates, price promotions, sell-through rates etc., and combine them with the stock keeping unit data of a product at a certain location and time. This data is assessed to optimize the local assortments to the individual customer segments in those locations.

Combining and connecting e-commerce with in-store shopping is the best way to offer a seamless shopping experience to customers. Online platforms are now lowering entry barriers by providing affordable substratum to new entrants. Big commerce allows anyone to set up an online store in less than 30 minutes. Since its establishment in 2009, it has already processed 17 million orders for as many as 35000 clients.[3]

Retailers have made the process of shopping as simple as walking into a store, scanning an item and paying for it with a series of taps on a smart device. On the other hand, online buying allows buyers to select a product by making comparison with alternate vendors and doing a best online shopping. As we progress deeper into the future, the shopping experience is bound to change even more. New and improved technology will be a step below of trends and needs for the customer. It is up to the companies to keep up with these trends if they do not want to lose customers. They can do this by harnessing big data wherever available—be it through social media, surveys or just a plain old feedback.

Learning

Educational institutions sticking with traditional methodologies have been experiencing a decline in revenue. These institutions must implement digital learning that exceeds a set of new expectations.

The demand for knowledge continues to be outpacing supply. As the world's population grows, so does the number of students who enter tertiary education each year. The total number of students in higher education is expected to reach nearly 380 million by 2030, 472 million by 2035, and more than 594 million by 2040– all up from roughly 216 million as of 2016.[4] Providing education to such a large number of learners using traditional modes of delivery would require tremendous amount of funding and effort, namely building roughly 4 universities per week for the following 15 years.

In order to educate the students of the future, CX analytics should be put at the service of understanding the mentality of the student, ranging from their preferences to their needs. Thanks to data on course enrolments, graduation rates, grades and feedback as they help education providers to develop a much more in-depth understanding of each student's exigency. This will allow institutions to make changes to their learning techniques accordingly and consequently provide an amazing student experience. For instance, publishing giant, Pearson acquired a learning analytics startup to strengthen its personalized learning offerings.

Leading companies are already making the most of the latest digital tools to propagate, share/acquire knowledge. For instance, Bank of America recently partnered up with Khan Academy, a not-for-profit provider of online education courses, to offer online learning on finance. The United States Navy has done the same with Institute for the Future, a think tank, to launch a massive online gaming tool that will help those craft strategies to fight piracy.

Staying Healthy

Data is at the center of a new trend in healthcare industry. Medical technology is increasingly becoming connected to a wide range of secondary devices that allow medical professionals to treat patients with greater insights and efficiency. Cloud based platforms are improving the overall treatment for patients through increased transparency. Integrated data from app-based wearable monitoring devices enables personalized medical analysis. The explosion of data will not subside, as consumers continue to demand better information about their healthcare options, costs, short and long-term outcomes, as well as other potentially life-changing factors that go into their treatment.

The more tech-savvy consumers are using the numerous mental-health apps that connect people with licensed therapists and allow them to chat by text/even through a video call. It is cost and time efficient for both parties, thereby delivering an exemplary consumer experience.

Marjorie Bessel, the chief clinical officer at Banner Health, shares her view over the future of healthcare: *"I think what the next ten years will look like for the industry will be lots of electronic tools, apps, portals that help us engage with the patient and help the patient engage more in their own health, their own health outcomes and how they decide to get there."* All these effects put patients and consumers in the driver's seat, steering wheel in their own hands. They will be calling the shots, while the tech will make sure their shots are reaching their treatment goals better than ever.

Entertainment

The digital transformation disrupting traditional value chain is the key market driver behind the rise of CX in media and entertainment. Consumers expect digital content that serves to entertain and be delivered to all their devices in an engaging way anytime and every time.

Advertisements and subscriptions are two key revenue streams for the media industry, and both need to be balanced. Payment/subscription revenues are not yet working in digital for many traditional players as they continue to fail to focus on the customer. The key is to build models where customer pays for value and experience; they receive, not just "for the content." Let us expand on this idea a bit further.

Graeme Noseworthy, IBM's Senior Social Strategist, has the perfect analogy: *"Where we used to say that 'content is king', we can now argue that the consumer audience is king, and the content is the castle we must build around them."* Media companies with access to huge data sources are trying to build close relationship with customer and understand them on an individual level. They are now able to analyze customer and behavioral data (who they are and how they act) simultaneously, creating accurate, detailed and personalized customer profiles using this information and predictive analytics. Media companies can now recommend content in real time, and that content can be developed to have substance that truly connects with their audience.

Disney has set a benchmark in the entertainment industry when it comes to achieving perfection. The company extensively uses technology to improve their customer experience, beyond the limits of their legendary theme park. My Disney Experience; a website and an app that requires one to log in with their personal Disney account. This application was launched, so that Disney's customers could thoroughly plan their vacation online; make dinner reservations and other quick selections. It also helped the customers explore the park using an interactive map, filter activities and check real-time show times, buy photos of themselves that were taken at the park. The website provides information about all of Disney's attractions, accommodations, and things to do etc.

The management team should always make it their mission to focus on their customers. The employees, stakeholders and the overall ecosystem must be aligned to this paradigm shift of customer experience.

Travelling

IBM's Bruce Speechley—*"Technology is now enabling people to experience the vacation before they arrive."* Experiencing means to explore, and exploring online means leaving digital traces. Companies uses this data to provide better customer service to individuals. Here are a couple of examples.

Delta Air Lines for instance, through exhaustive customer segmentation analysis identified that 5% of their customers accounted for 26% of their revenue. They then employed what they termed the "listen/respond/listen model"—listen to what those 5% of customers want, respond with products and services according to the that information, then listen again to see if customers recognized the value of Delta's response to their needs.

On the other hand, hotel chains like the Ritz-Carlton Hotel Company are managing their customer database centrally but use predictive modeling and purchase third-party data. This helps them understand what the customer might do next and how they can target them with the right message at the right time, and ensure they are receiving a personalized Ritz-Carlton experience.

The disruptive effects of crowdsourcing companies like Airbnb and Uber, on the travel industry have already been discussed. Though, crowdsourcing is innately innovative and efficient, it has also helped the companies to inevitably focus on the customer experience. However, disruption by digital innovation can be easily traced across the travel industry. Today, no company is immune to the demands for transparency that are given with crowdsourcing. Consumers are drifting to online marketplaces to thoroughly review and compare their options. In today's world, all it takes is a minute for travelers to tell hundreds of people how exactly their trip went, either on social media/an online review site. Online feedback is the digital version of word of mouth and can make/break a company. Businesses these days must focus on providing an exceptional customer experience in order to set themselves apart. As the rate of customers using digital technologies continues to grow, companies should continue to employ blends of innovative ideas in order to meet the demanding needs of their customers.

Paying

Leaders in the payments industry are teaming up with organizations outside their industry to get the capabilities to improve customer experience. For instance, American Express customers can use their loyalty points to pay for taxi rides in New York. Thanks to the partnership between the Credit Card Company and VeriFone, a provider of point-of-sale technology. BNP Paribas an international banking group joined forces with Proximus, the largest telecommunications company in Belgium, to pilot a digital wallet application. Another prominent example of innovation in CX in the payments industry, is the development of pay-at-the table service in restaurants.

Organizations are trying to innovate new ways to provide better customer experience and simultaneously take their brand to the next level. Therefore, excellent CX is vital for ensuring repeated purchases, positive recommendations and overall brand loyalty.

A significant change has occurred in transaction banking due to the digital revolution, and its impact certainly goes beyond consumer payments and retail banking. Introduction to faster and more convenient payment options in retail have led customers to demand similar conveniences and levels of service from transaction banking. Transaction bankers have witnessed these unusual payment systems in consumer banking and are now aware of the upsetting menace looming over their comfort zones. As payment technology and customer preferences evolve, companies must choose a flexible and secure solution that will introduce alternate payment methods in the future.

3

Powering and Serving Digital Clients

"Technology makes the world a new place."
— **Shoshana Zuboff**

Digital is driving major shifts in the way we work, travel, play and live. This shift has created a drive for businesses to reshape the way they interact and transact with customers and vendors by making seamless connectivity, a reality. To successfully navigate through the digital era, companies are altering traditional processes by reinventing their businesses.

Today, there are over 7.7 billion people on earth, of which about 5.11 billion have mobile technology and over 4.39 billion connected devices.[5] These numbers are quickly exploding in this digital world. Mobile is one such media that helps people stay connected in the digital world.

Although digital transformation has grown beyond marketing, it was this marketing segment that first adopted digital technology. Digital has played a crucial role in enhancing the quality of customer experience and customer engagement. *"Digital customers continue to expand and explode businesses beyond marketing. New products, features and services are being offered to increasingly more customers. The demand from customers is leading businesses to rethink their entire operations, which is again possible in the first place because of digital."*—Mike Sutcliff, group chief executive of Accenture Digital.

The prime motivator behind digital transformation is improvised business outcomes. Implementing digital transformation and measuring its success can be highly challenging for organizations due to the persuasive nature of digital technologies. *"A successful digital transformation is an enterprise-wide effort that is best served by a leader with broad organizational purview. Market pressures are the leading drivers of digital transformation"–* Brian Solis, Principal Analyst, Altimeter.

Data analytics is recognized as another block of the foundation for a successful digital transformation. From a top to bottom approach, organizations develop a culture of digital DNA at its core. A bottom-up approach of investing is required to successfully incorporate digital technologies and capabilities across a business. Leaders should educate people to realize that technology is not something that will replace them in organizations but will enhance the scale and quality outcome of their efforts and talents. For instance, telecommunications are one of the areas where the ability to adopt new digital technology efficiently is necessary to survive in this competitive market. 'Sprint' focuses on large-scale data analysis. The company has adopted an open-source platform called Elastic Stack in order to search, sort and analyze large amount of data from a vast variety of sources like databases, emails and others. With this data, the company's IT team analyzes the area where customers face problem, when they are transacting online.

Innovation in business products and services is a team sport. Innovation and digital transformation efforts are inextricably linked at all levels. CEOs must grasp the idea that tech is neither an abstract idea nor a computer program/machine. They need to build an ecosystem that reaps the advantages of new technologies that can rapidly redefine how products and services are created and delivered.

Accenture is at the forefront by helping organizations in their digital transformations. Pierre Nanterme, the late and former Accenture Chairman, and CEO shared his view on importance of changing the mindset—*"I believe it's a time when you need to rethink, in a very profound way, almost everything, from your strategy, from your leadership, from your current mobile positioning, branding, and it's a great opportunity for all companies, including Accenture, to very significantly reinvent themselves. Given what's happening with this—let's call it a 'digital revolution', there is much more at stake today than probably in this last 20 or 30 years, and so the need for reinvention is greater, and the risk is as well, more important."* Nanterme and his leadership team aimed for success by concentrating on their large business divisions such as Accenture Strategy Accenture Consulting Accenture Digital Accenture Operations and Accenture Technology etc. Together, in Accenture's fiscal year 2019, the businesses represented USD 43.2 billion in revenue and were driven by 492,000 global workforces.[6]

The advertisement industry works on developing deep skills that are increasingly required in collecting and analyzing customer insights with a focus on adopting new technologies. This can help in personalized advertisement strategies that can reach out to audiences in a very active manner.

Realized in entirety through the Internet of Things, the trend of connecting people to people, device to device, machine to machine and almost everything to everything is on the rise and is encouraged. Companies have also been making solid decisions with predictive analytic algorithm development and cloud by either implementing software in the cloud/supporting clients moving to the cloud from their legacy systems.

Researching and Developing strategies for companies has changed dramatically. Digital technology has shrunken the response time of new product strategies to months, from years. There is no longer a 3-5 week wait—companies get answers to their most anticipated questions very quickly. Social networks have made this possible. Digital disruptions have ultimately resulted in the focus of much current research, rightfully revolving around how quickly a product can be marketed and, more importantly, how well it is going to serve customer needs in practice. We have entered into an era, where many traditional business theories will be re-examined and re-interpreted. *"In many respects, it is also about a different interpretation of the 80/20 principle. In that, you develop 80% of the market and allow others to invest resources in helping unlock the last 20% of the market which is proving challenging to access. Approaching markets and businesses in this way is a tricky proposition for companies to get right. You need to find a balance between current demands and future developments. This also needs to be done against the backdrop of digital evolution and in conjunction with a network of collaborative partners that can strengthen your position."*—Audrey Mothupi, chief executive of Systemic Logic.

The digital transformation of businesses is at the top of the C-suite agenda. It is indeed the time of merging and co-mingling roles of CXOs and the era of interconnectedness and collaboration. This is because businesses can now buy things as a service. Digital is being seen not just as a technology enabler, but as a booster to the client's business, that encircles operations at every level of the organization.

4

Open Innovation
- Guided and Predictable Transformation
Author: Avnish Sabharwal

"If you are not failing every now and again, it is a sign that you are not doing anything very innovative."
— **Woody Allen**

Digital has emerged as a key driver of differentiated and disruptive value creation. It has enabled low-cost digital startups to lock horns with well-entrenched industry players; for example, by offering financial services without being banks, taxi services without owning taxis and lodging without being hotels. Who could have imagined, for instance, Paytm emerging as one of the largest payment platforms in India so quickly, with the potential to disrupt traditional banks that have dominated the landscape for decades? A report by Innosight predicts that 75 percent of the companies listed on the S&P 500 Index, between the period 2002–12, will be replaced by tech-savvy companies by 2027.[7]

These new breeds of organizations are the future of commerce, nonprofits and even government. The secret to their success? They have built business models that can keep up with—as well as take advantage of—the ever-accelerating pace of technology-driven change that defines our times. As a result, they are scaling swiftly—at least 10 times faster than other companies competing in the same space. Some can go from a hundred to millions of customers in a remarkably short timeframe. We call them exponential organizations, or ExOs.

ExOs are transforming the marketplace and taking business away from large, long-established organizations. These digital disruptors are dissolving traditional boundaries between industries and accelerating the pace of innovation. Big incumbents are recognizing that they cannot keep pace by relying only on their internal innovation capabilities. Instead, they need to leverage the innovation ecosystem that has arisen with the democratization of entrepreneurship—which has fueled radical thinking and sparked ever-faster ideation and execution of fresh ideas.

In India alone, entrepreneurs are supported by a thriving ecosystem of more than 100 venture capital funds, 7000-plus angel investors and more than 350 incubators and accelerators. Gone are the days when funding was the number-one challenge for a startup based in this vast nation. In 2019, startups in India (including e-commerce platforms) received as much as US$12.5 billion in investments from Indian and foreign investors, a significant rise from the US$5 billion invested in 2014.

What's more, the pace of innovation—already brisk—is expected to accelerate even further. Accenture estimates a huge potential loss of growth opportunity—US$52 billion—for Indian companies that fail to forge ties with the right partner for digital. We believe that large, established Indian companies can still defend their market while also spurring exponential growth. How? By adopting open innovation to tap into the startup ecosystem most strongly aligned with their industry.

Breaking down corporate walls

Through open innovation, organizations draw on external technologies, solutions, knowledge and other resources early in their innovation efforts by forging an ecosystem that spans emerging technology companies, start-ups, Venture Capitalists and academic organizations. Many partners with a range of players in a global ecosystem to jointly develop new platforms and applications, enhance core offerings or expand into new markets. This approach enables companies to bring in numerous new ideas quickly that they can use to enhance their operations, including product development and process improvement. It thus helps save time and money.

Some companies are even sourcing game-changing ideas by forging mutually beneficial partnerships with other ecosystem players. For example, Philips and the Eindhoven University of Technology (TU/e) announced a strategic cooperation aimed at accelerating exploration and development of digital innovation in healthcare, lighting and data science. The two organizations have set up a program that will let researchers from both entities work together to develop new digital technology in real-life settings. Through this partnership, TU/e will add more than 70 PhD positions, of which Philips will finance 30. In total, more than 200 researchers, professors, doctoral candidates and students will be collaborating closely.

But these are rare examples. Most companies have instead invested in corporate research and development (R&D) and in initiatives centered on driving new growth by using their existing, internal resources. They seldom engage with elements of the broader ecosystem beyond their corporate boundaries to support their internal efforts. For those that do take part in open innovation activities, some are ad hoc. That is, the company notices an opportunity and reacts to it, rather than taking a more proactive, systematic approach.

Adopting open innovation is a must for established companies if they want to continuously harvest innovation, but the model does pose challenges. Cultural differences, technology risks, security threats and difficulty scaling an innovation can make it difficult for organizations to realize open innovation's full potential. To get the most from this approach, businesses need to mitigate the risks and ensure that innovation efforts align with their corporate vision and support their business priorities. They also must manage the complex, time-consuming task of building and maintaining relationships with many partners, including startups—given that the quality of such relationships can make or break open innovation efforts.

To master all of this, established companies need to excel at five practices: define the innovation strategy, find the right partners, select a winning formula, deploy innovation at scale and foster an open innovation culture:

- **Define the innovation strategy:** Clarify the business goals you want to achieve through open innovation, by identifying opportunities and threats arising in the market and assessing how digital disruption in other industries could help your enterprise seize those opportunities or combat the threats. Corporates could opt for "Target Innovation," which focuses on a specific business problem such as an omnichannel strategy in retail or payments in banking, or they could opt for "Disruptive Innovation," which builds a new business model for driving digital revenues.

- **Find the right partners:** Successful open innovation relies on an organization's ability to identify the right set of partners. The process requires sifting through multiple technology innovations and identifying partners that can help you create more valuable solutions than your company would have developed on its own. You can shoulder the unenviable task of identifying and managing multiple partners on your own while also running your day-to-day business. Or you can work with a bridgemaker.

Bridgemakers:

  - o Serve as intermediaries to help connect an organization with appropriate partners.
  - o Act as a buffer between partners that have conflicting cultures.
  - o Provide support in mitigating the risks inherent in open innovation.
  - o Help organizations pilot and deploy innovative technologies.

Accenture Ventures and Open Innovation serves as such a bridge maker for its clients which represent the G2000. It worked with a large bank to define the digital-banking ecosystem and to identify startup players that could help the bank accelerate its digital transformation program. Drawing on Accenture's experience with payments and analytics from other industries, we developed a solution for the financial services industry that helped the bank establish new digital operations and develop digital services and products.

- **Select a winning formula:** There is no one-size-fits-all way to adopt open innovation, however there are best practices and models to repurpose or adapt. You can select from an array of formulas or even combine them in ways that best meet your business objectives. Consider these examples:
  - *Corporate Accelerator:* Large incumbents engage with a group of startups on an identified innovation theme. Key examples in India include the Reliance GenNext Innovation Hub as well as the Target Accelerator Program, which has been launched by Target, a large retailer in North American with a global in-house center in India.
  - *Co-promotion:* A group of like-minded organizations come together to run an accelerator program. To illustrate, Accenture, along with the Partnership Fund for New York City, runs FinTech Innovation Lab. The annual 12-week accelerator program brings together early-stage financial technology companies and the world's leading banks. The program's objective is to give early and growth-stage companies the platform they need to develop, trial and prove their proposition alongside the world's leading banks. The program runs in New York, London, Hong Kong and Dublin.
  - *Exclusive Partner:* An organization teams up with one partner in an exclusive arrangement centered on implementing a specific innovative idea. For example, Accenture teamed with Isansys, a UK based IoT startup, to develop the iDOC physiological wireless monitoring system for healthcare. The device monitors a patient's electrocardiogram readings and transmits them wirelessly through a mobile device to a doctor or hospital, where healthcare providers can analyze the real-time data to assess risks to the patient.

- **Deploy innovation at scale:** Most organizations and startups have no difficulty working on a pilot together. But they struggle to scale the innovative solution—across the enterprise or to their customer base—because of several reasons such as lack of capabilities as well as the expertise to integrate multiple enterprise systems. To overcome these challenges, corporates need a clear business case, strong internal sponsor, the right partner and an open innovation culture.

- **Foster an open innovation culture:** Open innovation requires a different mindset about how innovation should be driven within an organization. At the most fundamental level, corporate leaders must foster a culture that embraces experimentation and open innovation and backs it up with investments of time and money. Examples include designing incentives and rewards encouraging employees to share knowledge and technology with external parties as well as easing the legal and procurement processes to quickly on-board startups.

Driven by disruptive technologies, the pace of change in the marketplace is accelerating. Traditional companies cannot keep up by relying only on their internal innovation machinery. Instead, they need to augment their technological capabilities with open innovation to compete against Internet-enabled startups as well as digitally savvy large companies. Applying the practices described above can help them manage the challenges that come with adopting open innovation—so they can extract maximum value from this powerful approach.

What are the Future Open Innovation Trends?

Almost 70% of the Fortune 500 enterprises have implemented some version of the Open Innovation model in their organization – from co-creation and partnerships to hackathons and crowdsourcing. However, the concept of Open Innovation has been evolving since the time it was first coined almost 17 years back by Henry

Chesbrough. Below we look at some of the new trends emerging in this space:

1. **More open to Open Innovation:** Many industries like Manufacturing and Pharma have been relatively slow to adopt Open Innovation practices compared to the ones which have seen massive disruption already like Retail, Financial Services and Media. The early adopters in India for example have been MNCs, Banks and companies and Tech companies like Microsoft, Cisco, google, SAP and IBM. This trend has slowly started changing and there is a healthy appetite now among Indian corporates across industries like Manufacturing, Construction, Pharma, Oil and Gas, Healthcare, Telecom and Consumer Products. A large industrial equipment maker based in India collaborated with a niche Internet of Things startup to create a connected boiler. The connected asset is helping the equipment maker transform its maintenance and support model in the short term and its business model in the long term. GE Open Innovation Manifesto focuses on the collaboration between experts and entrepreneurs from everywhere to share ideas and solve problems. One of GE's project is the First Build, a co-create collaboration platform, which connects designers, engineers, and thinkers to share ideas with other members who can discuss it together.

2. **Look East:** 2019 Global startup ecosystem report described that there Will Be No "Next Silicon Valley" instead, there will be 30 "next" hubs, distributed around the world, reaching critical mass driven by either regional as Singapore in Southeast Asia or Sub-Sector leadership (e.g., San Diego in Life Sciences. While none of them will be as big as Silicon Valley, each will thrive. For organizations starting out on their open innovation journeys, 'start-up safaris' to Silicon Valley and Israel have been an integral part of the initial discovery process. While these global startup hubs will still host a fair share of safaris, more organizations will start looking east - especially at China. China is making its presence felt in the global startup ecosystem as a hub for deep tech startups in areas ranging from artificial intelligence,

computer vision and electric vehicles to drones and genomics. Expect more China 'safaris' and bridge programs in 2020 and beyond. Singapore an Hongkong have emerged as major FinTech Hubs and India is home to the 3rd largest number of Unicorns.

3. **Accelerators 2.0 - Distributed, Virtual and Rolling:** Accelerator programs have been the low hanging fruit for many corporate open innovation programs, and they will continue to be so in 2020 as well. However, the format will increasingly move to virtual and hybrid programs where a cool working space is no longer the center of attraction. Also, the fixed term cohort model will increasingly give way to rolling cohorts based on real time business needs. Finally, more and more organization are reinforcing their open innovation efforts in different ecosystems by collaborating with different cities and keeping track of the entrepreneur's evolution. The Coca-Cola Accelerator program aims to help start-ups in eight cities around the world, Sydney, Buenos Aires, Rio de Janeiro, Berlin, Singapore, Istanbul, San Francisco, and Bangalore. Those start-ups aim to think in innovative ways to build the Happiness Coca-Cola brand. Samsung's accelerator program invites the collaboration of diverse startups across New York, Palo Alto, and San Francisco.

4. **From use-cases to broader business problems aligned with Digital Transformation:** In 2020 and beyond, Open Innovation programs at corporates will focus more on broader business problems and Digital transformation than peripheral use-cases as the focus shifts from tactical, one-off problem statements to scaled deployments. Expect Open Innovation to become a key catalyst for Digital transformation initiatives of large enterprises. Bangalore International Airport has a comprehensive Open Innovation agenda, collaborating with multiple digital startups all the way from Customer experience to energy optimization as part of their Digital Transformation agenda.

5. **Investments and Acquisitions:** Organizations who have already experimented with open innovation models like

accelerators, hackathons, crowd-sourcing and co-innovation programs will show more appetite for investments and acquisitions in 2020. This is especially true for industries which are more vulnerable to disruption where incumbents look to tap new markets and open new revenue channels through these non-organic bets. Expect a surge in Corporate Venture Capital beyond front-runners like Intel, Facebook, Google and Reliance.

6. **Partner diversification:** In the last couple of years, Open innovation programs have mostly focused on technology enabled startups in the B2B and B2C space. 2020 will see more interest from organizations in engaging with start-ups as well as academic institutions and research organizations which specialize in deep sciences as well as tech for good. In order to build a mathematical algorithm that can determine the optimal content of medical kits for NASA's future manned missions, NASA collaborated with TopCoder, Harvard Business School, and London Business School. In this collaboration, TopCoder members provided 2,833 code submissions to help NASA build the intended algorithm. Top solutions were provided from UK, Japan, Indonesia, and Brazil.

7. **Embracing Bridgemakers:** Accenture research found that incumbents and startups together spent US$3.2 trillion on innovation-related initiatives between 2012-2017. This included investments in research and development, technology, corporate venture capital, and mergers and acquisitions. However, only 14 percent of those surveyed turned these investments into real value. As there are growing concerns about financial sustainability and how to measure the return of the investment of these programs, collaborating with Bridgemakers might grow as more incumbents are moving away from internal management to third-party managed accelerators. In future we will witness more organizations embracing bridgemakers like Accenture Ventures or Microsoft for Startups as strategic partners to give arms and legs to their open innovation programs and

help reduce time to value by moving from PoCs to large scale deployments.

To conclude, moving forward, corporates will move several steps further in their journey to leverage the power of Open innovation, especially by experimenting with multiple Open Innovation models, collaborating with a wider set of ecosystem partners across Geographies and thinking deeply about realizing real value from their investments by working with bridgemakers.

There are many ways to create lasting partnerships leveraging the Open Innovation model that deliver long term value. The key is to recognize that the best partnerships for a business may not always be conventional, and or seem obvious. But if invested in the right manner, they unlock new opportunities that pivot on your core strengths to tap new markets and make greater gains. Driving innovation and making it sustainable is not a "one and done" exercise. It's a continuous evolution that requires business leaders to take cognizance of their organization's appetite for taking risk and coping with failure, address 'Not invented Here' syndrome through cultural change and considering your open innovation partners as an extension of your organization.

3.1 Intelligent Enterprises

Quick thinking allows quick decision making. Decisions need to be backed up with accurately analyzed data and must be fast enough to beat the competition. The art is to balance, and there is no balance without strong footing.

Today, LinkedIn is the largest professional social networking platform, on a global level. It offers a wide variety of business-oriented services integrated in a single platform. It is boosting its business sales by utilizing the user and prospect data. Over the past few years, LinkedIn's revenue has steadily grown, with over USD 6.8 billion in 2019.

Data has undoubtedly been proven to be one of the key driving factors of business decisions across industries. Modern startups are targeting this niche of data collection to provide industry leaders with data analysis tools, designed to make them better at what they do. Consider the example of Saama Technologies, which used the Partner@Speed program to attain world-wide support from prime big data and analytics companies like Informatica, Hyosung, Cisco, Tableau and Salesforce. Saama employs its Fluid Analytics engine accelerators in combination with their industry-specific consulting services. They provide big data solutions relevant to the world's largest public companies. Data and analytics are speeding up the outcomes, delivering transformational business for world-class clients, and changing the game.

Mark Ledbetter, former Senior VP of Solutions Engineering at Hortonworks, sums up this need for speed, without compromising on accuracy, and shares his company's solution: *"Our customers want insights delivered at the speed of business. The software's that we use in integration with other companies with a similar strategy has helped us accelerate customer success with reduced implementation times."* The digital age is infinitely more about sharing, than it seems at first glance, and sharing promotes efficiency in almost all contexts. Smart factories are an example of how a business concept with long-established connotations of rigidity can have a bright future.

Smart Factories

"The factory of the future—the smart factory—is a paradise of efficiency where, defect and downtime, waste and waiting are long forgotten issues of a long-forgotten age." — Industry Week. It is predicted that the newly established factories will have plant managers and CIOs crafting together, a solution that will seamlessly incorporate data with production. This helps them to stay on top of every movement of every gear, in order to deliver services like never before. Combining advanced tools with high-tech workers will result in an optimal assimilation of manufacturing and technological advancement. Investigating the 'Siemens Gadgets Works Office' in Amberg, Germany, feels like walking around tech heaven, with an area of 108,000 square feet, it is striking in its level of technological advancement. The company's intelligent machines are used to coordinate the factory production and worldwide distribution of Simatic Controllers. This includes a custom made-to-order process,

involving more than 1.6 billion parts for more than 50,000 annual product variations. Siemens assembles around 10,000 materials from 250 suppliers to produce 950 different products in the plant.[8]

The introduction of smart factory technology will improve the reliability and flexibility of advanced automated machines, thus changing assembly line production as well. This cuts down the costs associated with changing production lines and machine set up. The smart factory will enable manufacturers to make a wide variety of products and offer customers the products that are more customizable. Being capable of manufacturing a smaller number of products at mass production price will give smart manufacturers significant competitive advantage over their traditional counterparts. *"The intelligent networking of industrial devices promises to deliver productivity gains, but not all manufacturers are ready to make the leap to the smart factory. Connecting efficiency and productivity is the key for the future of the industrial sector,"* entrepreneur Arthur Wisser.

The term "Industry 4.0" sneaked into use. It first appeared at the Hannover Industry Fair and was the focus of discussion at the 2014 exhibition. Industry 4.0 has digitized the manufacturing sector and helped in fostering the smart factories. Being the fourth industrial revolution after mechanization, mass production and digitalization, the focus is on exchanging data and automation using digital technologies. Over the past few years, the manufacturing sector saw a lot of IT-integrated products put onto factory floors, such as autonomous industrial vehicles, and the same happened with production processes. This integration included computer aided design, enterprise resource management and planning. Smart factories are characterized by adaptability, ergonomics and resource efficiency. One novelty such factories bring to the table is the inclusion of customers and business partners in the business and value management process.

Integration, along with sharing, is a common theme in the digital age. Helmuth Ludwig, the CIO of Siemens North America's Industry Sector, emphasizes that *"The future of smart manufacturing is today. Previously, the industrial value chain including product design, production planning, production engineering, production execution and services were implemented separately. Today, new technologies are bringing these worlds together in exciting ways. "* The fact that made the 'Amberg Siemens plant' successful, was combining three specific and critical technologies, namely Product Lifecycle Management (PLM), Manufacturing Execution System (MES) and Industrial Automation. Whether a company decides to produce a modern commercial aircraft, fuel efficient cars or high-performance golf clubs, the technology used in PLM, MES and Industrial Automation are helping manufacturers to realize and achieve top-line growth by increasing productivity and minimizing risk.

Connected Supply Chain

For companies and businesses aiming to deliver products to their customers on time and without compromising on quality, an efficient supply chain is a must. In this ever-evolving global marketplace, managers are required to employ more innovative and proactive strategic approach, to be able to optimize the supply chain and reduce costs throughout the product life cycle. The supply chain management strategies need to change with each phase of the product life cycle—Launch phase, Growth phase, Maturity phase and Market Decline phase.

Communication and information are vital in maintaining a seamless supply chain. Providing a proper access without compromising on security can help organizations support an incredible workflow, that promotes innovation and eases collaboration. As an example, CEMEX faced high transportation costs and spoilage as customers repeatedly changed their orders and delivery schedules. Using global positioning system (GPS) sensors, mounted on cement trucks and linked to a central control center, CEMEX can now reroute trucks dynamically, based on up-to-the-minute information about changing customer requirements. As a result, CEMEX reduced delivery time from three hours to 20 minutes, cut the number of delivery trucks by 35%, trimmed operating costs by USD 100 million and improved on-time delivery.[9]

Supply chain performance can be greatly improved through innovation. Technology and innovation play major roles in cutting down production costs by designing products that are easy to manufacture. The assembling process can be optimized by designing the product to contain a minimum number of subcomponents/designing subcomponents that are easy to assemble.

In order to enable an intelligent, connected supply chain, a platform that enables a deeper visibility across the value chain must be in place, providing a comprehensive look at health, risk (always a crucial factor in decision-making) and profitability. The platform should securely manage, automate and supervise the complex network of systems and things that need to access resources and share them. Collaboration among various people is key in this context. The platform must allow them to seamlessly connect with each other by delivering the right information to the right entity at the right time. Finally, the platform can be on cloud, in order to reduce cost, help simplify infrastructure management and accelerate time to market.

This is what "Internet of Things" (IoT) is, and has a huge impact on connected supply chain management. As per Cisco, IDC and Gartner reports, a notable increase in the number of devices making up the IoT will have an acute impact on how supply chains operate in the future. Cisco` predicts that the number of connected devices on the Internet will exceed 50 billion by 2020 on a global level, i.e. 6.5 devices per person.[10]

The impact of IoT-supported supply chains is wide and significant. IoT has already incorporated a number of commercial operations like telematics that are used in trucking fleets to improve logistics efficiency.

Here is a piece of wisdom from Accenture that emphasizes the inevitability of transformation, if growth is to be achieved. *"Every high performing supply chain is a digital supply chain."* In this age of digital disruption, every traditional business is transforming into a digital business to stay in business. Digital disruption has especially impacted supply chain management, and no business can reap the optimal benefits of digitalization without reinventing their supply chain strategy. To successfully transform the supply chain, the businesses should evolve into a digital network, connecting the physical flow of materials, talent, information and finance. The supply of the new age can be characterized as more connected, efficient, scalable, and intelligent than the traditional one.

Intelligent Infrastructure

Today's public utilities face increasing and often conflicting pressures to restrain costs and comply with new mandates. More than ever, companies need to use their infrastructure wisely. In Siemens, they know and take advantage of the fact as they say, *"IT and automation are expanding the potential of infrastructure across the world. Solutions for sustainable power distribution, efficient traffic systems and efficient intelligent buildings are becoming more flexible and adaptable to new conditions."*

An integrated digital infrastructure plays an important role in the development of modern cities. Smart transportation, intelligent buildings, smart and independent power grids are their basic characteristics. An intelligent infrastructure enhances the potential of any location to become the location of choice for digital businesses.

"In a fully *realized intelligent infrastructure, the desktop will be provisioned based on the unique user profile, providing the applications and infrastructure services each employee needs to support his/her unique role,"*—the Wall Street Journal. The infrastructure will be able to determine the employee's location— office, mobile/home, and provide user experience accordingly, to maximize workplace capabilities. This infrastructure will ensure that devices are always on and are connected securely to a network at required environments. The current technology is not quite ready to bring the vision of an intelligent infrastructure entirely into reality. The IT leaders should start preparing for it.

While designing intelligent infrastructure, the focus should be on the end user. It has now become vital for companies to provide its users with services, whenever and wherever the users demand. IT can unlock workflows that meet the needs and expectations of the ever-needier end users of the new age. For example, user expectation on the rise is having a "user app" allowing access to all the business applications on their mobile devices. While these applications must work at the same level of efficiency and in the same manner irrespective of device. Companies must work closely with developers, building apps and need to pay detailed attention to the user interface and overall quality of user apps. It is pertinent to embrace the 'workplace-as-a-service'. Cloud services are quickly gaining popularity for non-core operations, as they offer a low-cost alternative.

Change is always difficult, but it leads to long-term possibilities for growth as opposed to stagnation first, and then drowning in the sea of innovative startups. Businesses that went through change are now riding this new tech wave. Once the business has moved over to the new systems and infrastructure, processes will get smoother, leading to better decision-making and thus perpetuating progress.

Smart Product Lifecycle Management (PLM)

The process of managing entire lifecycle of a product, encompassing everything from initiation, through engineering design, manufacture, and disposal to service. An organization's product data is one of its most valuable assets. Different levels in an organization have various roles where they evaluate product data to make decisions that greatly affect the business in various ways.

The adoption of a new enterprise management solution requires weeks of intensive training. The daily active user needs to understand every minute detail about operating the software and filtering the data that the software manages. While this training is ideal for people who manage such software on a daily basis, it is an overload of data for other stakeholders. Smart PLM solutions provide the access of information to stakeholders/anyone who is not a regular user of the software. For instance, if someone in manufacturing, sales, marketing/procurement needs product data occasionally and lack training for more complex PLM tasks, they can simply use a role-based app to get specific pieces of the information.

By incorporating elements of smart PLM technology, the complexity of global product development and manufacturing is drastically minimized. Product development depends on valuable feedback from both within and outside the confines of a business. The brand-new PLM generation brings socially-oriented, collaborative techniques, which pave the way for advanced social media strategies.

Search Manufacturing ERP provides an example to showcase the comprehensiveness of PLM—*"A company developing a new washing machine could capture ideas and requirements in PLM. They could use those to develop some conceptual designs and collaborate on them with the marketing department. Then, they could manage new product development project through the design of all the related specifications, components, software, documentation and other deliverables required to launch a successful product. Ideally, they would manage manufacturing and service processes with PLM as well."*

Intelligent Services

According to Harbor Research, there is a significant sluggishness across businesses: *"It is interesting to compare consumer focused smart connected business models to industrial B2B models. The industrial players are moving so slowly to evolve their business model designs, that they risk implementing solution concepts developed in the late 1990's by about 2020."*

Digital giants like Apple, Google, Facebook and Amazon are setting examples for companies that are developing a smart service business model by combining technologies from different domains, and in turn providing services and solutions in the way consumers want.

A company wanting to become a player in the smart services arena should make a strategic decision about the role it wants to play in the overall 'Business Ecosystem'. Monitoring and controlling the data entry point to a platform is a significant part of the new, smart services. Suppliers of consumer demands and data interfaces of smart, networked products and services will lead the way. Such companies will also seek to grow and further increase scalability by creating digital ecosystems.

According to Harvard Business Review, *"Smart services are a wholly different animal from the service offerings of the past. To begin with, they* are *fundamentally preemptive rather than reactive/even proactive. Preemptive means your actions are based upon hard field intelligence. Smart services are thus based upon actual evidence that a machine is about to fail, that a customer's supply of consumables is about to be depleted, that a shipment of materials has been delayed and so on."*

Smart services can eliminate all kinds of unpleasant customer experiences. Thanks to its data-based intelligence and consumer behavior, that enables manufacturers to gain unprecedented R&D feedback and insight into customers' needs, and thereby providing even greater ongoing value.

Machine intelligence has reduced the issues arising from tasking humans with gathering real-time data. Devices running on machine intelligence can digest billions of data points. Processing and controlling it, depending on the data itself. This helps the decision-makers by providing them with more visibility and insights pertaining to the business's assets, costs and liabilities whenever needed.

Companies like GE, Siemens, Honeywell and ABB etc., are the leading smart service providers. Honeywell's aerospace, GE's jet engines, ABB's power plant equipment and Siemens's medical equipment, locomotive, all produce assets of critical value to customers. They are using various kinds of networking to carry out remote monitoring and diagnostic procedures.

The innovation paradigm has changed from closed to open innovation, providing a significant boost to the industrial research and innovation processes. Open Innovation contributes and improves one's entrepreneurial knowledge. Industrial sectors are keen to exploit and access knowledge outside their boundaries to accelerate their innovation processes.

5

Intelligent Automation
- The driver for Digital Revolutions
Author: Frank Pendle

"Speed is the new currency of Business."
— **Marc Benioff, CEO of Sales force**

Intelligent Automation is defined as the orchestration of several new and emerging technologies. The coordinated use of these technologies is being used today to drive the next quantum leap in business value. This business value emerges in different areas such as enhanced efficiency, increased worker performance and satisfaction, reduction of operational risks, and improved response times and customer journey experiences. These technologies can be grouped in categories such as:

1. Advanced Analytics: Companies are producing vast amount of data, and over 70% of this data is not used in any way[11]. New systems and algorithms are able to process, identify, structure, and present data that drive actionable insights.
2. Artificial Intelligence: Applications that use human skills such as visual perception, speech recognition, recognition of printed and hand-written text, decision making, and language translation[12].
3. Machine Learning: The application of algorithms, statistical models and logical processes to analyze situations and make decisions without using explicit instructions. This differs from "standard" Artificial Intelligence because ML systems

change and improve over time, as opposed to be a static solution.

4. Robotic Process Automation: Tools that automate tasks that are repetitive, rule-based, and/or manual. They work by replicating the actions of an actual human interacting with one or more software applications to perform tasks such as data entry, process standard transactions, or respond to simple customer service queries[13]. The use of RPA allows humans to concentrate on more value-added components of the value chain, while vastly improving speed, quality, transparency and accountability within existing processes.

5. Business Process Re-Engineering (BPR): The analysis, design, optimization, and deployment of strategies to support an organization's objectives by continuously improving process output, flow, satisfaction, cost and effort. Processes in and between organizations are benchmarked throughout several different categories, and then continuously monitored, modified, and optimized. The main difference between this technology and the classical "Hammer" process reengineering is that contemporary BPR can be largely automated.

There is tremendous impact being created across enterprises. Some of these examples include[14]:

6. 20-35 percent annual run-rate cost efficiencies in large-scale processes such as Order to Cash, Record to Report, Hire to Retire, Procure to Pay, etc.

7. Reduction in straight-through process time of 50 to 60 percent.

8. ROI most often is in triple-digit percentages.

Paradoxically, many of today's tactical automation implementations are not successful in delivering the promised business value. In fact, several sources cite a success rate of less than 50%[15]. So how can we reconcile the radical rethinking of the large-scale processes' companies use today with these new and untested capabilities?

The answer lies in the intelligent, focused application of these next-generation tools. There are three main value clusters for Intelligent Automation being used today:

9. Augmenting human capabilities: "taking the robot out of the human"[16], i.e. taking over repetitive, boring, difficult or dangerous tasks that don't add business value and lower client and employee satisfaction.

10. Orchestrating data fusion and application extensibility: data and applications are becoming consistently more complex over time. Creating interfaces between systems—some just created this year, other legacy instances that may be over 50 years old—is increasingly complex as well. Intelligent Automation enables fast, inexpensive and powerful interoperability between all data and applications. Moreover, these integrations occur at a fraction of the time and cost as compared to older technologies such as APIs.

11. Enabling new business models and tech capabilities: the effective use of Intelligent Automation enables new ways of doing business and achieving business value that weren't possible even a couple of years ago. This includes the virtually unlimited availability of skilled laborers, computing power, and sophisticated analytics driving actionable insights.

The dramatic potential that Intelligent Automation offers has certainly been noticed by the market. AI startups alone have mushroomed from near zero in 2000 to more than 12,000 today[17]. AI investment alone has reached more than USD 50b/year in 2019, with a 20% CAGR expected until 2025[18]. Most AI is developed by governments, academia or small companies, as the state of the art of the technology overwhelmingly enables small, tactical AI solutions (Artificial Narrow Intelligence—"ANI"). Artificial General Intelligence ("AGI") is much more difficult, so there are comparatively few large corporations that are working on these—mostly Alphabet (Google), Microsoft, Facebook, and the like. In these, though, the R&D investment is enormous—Microsoft alone is spending USD 8b in 2019 on ANI, not counting acquisitions such as the USD 1b Open AI deal. AI lends itself to consolidated financial plays, so AI-centric funds are growing over 35%/year (such as Softbank's Vision and Vision 2 Funds—this last, an AI-focused second $108 billion Vision Fund with LPs including Microsoft, Apple and Foxconn[19].

RPA has lent itself to much more traditional development, even though the hype cycle is no less strong than that of Artificial Intelligence. In fact, RPA grew at a blistering 63%/y pace in 2018, becoming the fastest-growing technology on the planet[20]. This is remarkable, especially since the total market is only about USD 1b in 2019. This remarkable growth has led to mind-bending results and valuations[21].

RPA solution providers like Automation Anywhere, Blue Prism and UI Path, have shown dizzying growth trajectories and high valuations that are unlike any other technology companies— already a vertical that lends itself to financial hyperbole.

This revolutionary, new way of reshaping business processes crafts, deploys, and manages new and existing value streams—and continuously improves them over time. These new value streams drive revolutionary operating models that leverage them to drive improved and new ways of doing business. Used correctly, Intelligent Automation is a powerful catalyst for change. It enables much higher speed to value, process quality, business agility, and lower risk and cost, while lowering risk and cost in ways not seen since the inception of Business Process Outsourcing.

Use Case 1

One example of Intelligent Automation is in rethinking Enterprise Asset Management for a major Oil & Gas multinational. This company has over 40,000 kms of pipeline in many countries. Pipeline Maintenance, Repair and Operations is very complex, expensive, and dangerous; typically, crews are sent out to physically inspect the assets from time to time and perform repairs on them as required. This is a major issue: problems, unlike crews, do not operate on a predictable schedule. Further, when crews are on their maintenance trips, they do not carry every tool and part they may conceivably need. This inevitably leads to even more delay in resolving very serious issues. The most important flaw in the existing model, however, is in Environmental Health and Safety risks. Pipeline go through some nasty places; places that are too wet, too high, too underground, or too war-torn to be safe. In fact, 20–30 crew members die every year performing this maintenance.

This is a critical issue for the company. It costs USD 250 million/year to maintain, yet they have poor visibility into their asset performance, don't have enough personnel to properly maintain the equipment, and yet are exposed to the nightmare risk of people dying and disastrous spillages that are environmental, economic, and PR nightmares.

But what if? What if this Oil & Gas company could have unlimited eyes, ears and hands watching the pipelines every hour of every day? What if these new employees were trained to identify exactly what issues are occurring and could decide exactly what had to be done, and take the necessary steps to resolve the issue as quickly, effectively and safely as possible?

Of course, the hypothetical above is impossible using humans and the existing, human-centric processes. But, when we use the guiding principles of Intelligent Automation, the problem becomes solvable. The company is using hundreds of RPA Digital Workers (DW) to constantly watch pipelines through a sensor fusion of security cameras, drones, satellite imagery, and even geo-tagged social media. This information is consolidated into a data lake, which then is analyzed by several different kinds of Artificial

Intelligence. These AIs can be general or specialized, but each is responsible to detect types of issues. One may be specialized in detecting unauthorized personnel close to the equipment; another may be specialized in types and degrees of corrosion. These AIs, orchestrated by the DW, analyze and discuss the enormous amount of information they receive every moment—and then decide what to do about it. Any issue has a resolution workflow attached to it; the AI decides the appropriate resolution and the DW quickly connect to all the required data, applications and systems to execute on the solution.

Now, what if there's an issue the AIs have not yet encountered? If the AIs don't agree on an issue, they can bring in a human in the loop to solve the issue and teach them. An interesting example was when the solution first encountered graffiti on a pipeline. The IA solution could not find a process that covered the issue, so the DW sent an email to the responsible manager asking how to categorize the issue. The manager indicated the correct resolution process and emailed that back to the DW. The DW retrained each AI individually to recognize the new issue, and further retrained each DW how to process it.

The benefits of the solutions go beyond the detection and disposition of thousands of issues—following best practices, not making mistakes, and dramatically improving response and resolution times. The DW also interface with hundreds of legacy applications and data throughout many of the client's, third party contractor, and supplier environments. This allows great business agility and transparency at low cost—for example, an analytics system originally quoted at USD 5 million- and six-months' development time was delivered in three weeks and a cost of USD 200,000 using Intelligent Automation.

It's clear that the use of any one of these technologies by itself would not have delivered a fraction of the value that of their harmonized use. It is a testament to the power of this new way to deliver business value that Intelligent Automation is saving the company USD 80 million/year in lower maintenance costs and better outcomes. More importantly, it's saving lives as well.

Use Case 2

A major Manufacturing company used Intelligent Automation to lower its client and supplier relationship costs by over 90%. This particular company has an enormous number of SKUs and serves over 22,000 clients all over the world. Their products are complex, so they have over 50,000 suppliers. Clearly, the requirement to have a fast, reliable, and cost-effective supply chain is critical for this company, its suppliers, and its clients.

Typical supply chains are still heavily dependent on people. They need people communicating with shippers, preparing paperwork, ensuring compliance and accuracy. Further, companies have different application stacks and inefficient analytics that make it very difficult to get a picture of the company's supply chain performance—much less keep it running at peak efficiency.

But what if? What if you could have DW watching every single shipment—its products, paperwork, transportation, and client needs—every hour of every day? Further, what if you could create a super-smart DW that knew everything about your company—its applications, data, policies, best practices, etc.—and could make smart decisions based on the best data? Further, what if this DW could go to each supplier and client and connect to their applications and data to transact, process, and negotiate in the company's behalf?

In fact, the developed solution uses a Digital Worker that is given all the tools it requires to know how to operate on behalf of the company. Further, the DW has an AI component that understands how to connect with the supplier or client systems—and does so without human help. Machine Learning is used to find the best way to serve the company's interests; for example, it can detect the inventory for a certain supply is low and reach out to every supplier to ask for a quote, then negotiate and close without human interference. From there, the solution learns how to build cost-effective routing considering many variables (cost, time of travel, temperature, shipper quality, etc.) at the same time. DW keeps a constant watch on the shipment, and can auto-remediate a potential issue (for example, mismatching shipment documentation) without human intervention as well.

The results are obvious and dramatic. The company significantly reduced its reliance on humans to maintain its supply chain; it also improved the quality, speed, and agility, while gaining complete transparency and visibility of its operations. Another interesting result was that the company reduced the time and effort for its clients and suppliers to do business with them, providing opportunities for the company to get discounts and other advantages.

These use cases illustrate the power of strategically-placed Intelligent Automation. This ability to change the way companies do business has not been ignored by large corporations, which are scrambling to build the integrated capabilities necessary to build the next quantum leap in business value. For example, SAP has significantly augmented its automation capabilities in the last two years: not only has it improved its native Leonardo automation solution, but it also bought an RPA firm called Contextor in 2018[22]. SAP is offering Intelligent Automation skills to augment its own product suite, as well as extend processing and integration beyond its own products.

Other companies have followed suit, notably Microsoft with its October 2019 announcement of an RPA suite called Power Automate[23]. Growing out of Microsoft's Business Process Management solution called Power Platform (formerly known as Flow), Power Automate offers native BI functionality, low code and workflow management skills. It also comes with 275 prebuilt connectors for apps and services. As would be expected, it also has deep integrations with Office 365, Dynamics 365 and Azure.

Other tech companies are using Intelligent Automation to grow share of mind (and pocket). Appian, for example, uses RPA to take actions and coordinate the performance of its BPM software. 24/7 AI, a major Natural Language Processing and Decisioning solution, uses RPA and ML to create human-like "chatbots" that offer human-like interaction to call center clients. These "chatbots" come at a fraction of the cost of classical call center operations, making call center agents more than three times as effective.

Market analysts confidently expect other technology giants such as IBM and Oracle to join the fray very quickly. All major players recognize that they are playing catch up in the Intelligent Automation market—and that using IA strategically is critical for survival for both clients and themselves.

Intelligent Automation has the potential of forever changing business as we know it. Its potential to replace large parts of human labor is as game changing as the steam engine was for the Industrial Revolution. This change will not occur by itself, though; it requires a concerted effort by academia, corporations, clients, and thinkers that will define how to put these new components together to transform processes that haven't changed materially for a hundred years. The prize, though, is worth it: a world where humans can concentrate on being human—and let the machines do the drudge work.

6

Building a Digital Workforce
- AI, Cognitive Computing and Digital Workers

"The best way to predict the future is to invest in it."
— **Alan Kay**

The arrival of robots in today's companies is an unstoppable trend. According to Bill Gates, *"Robots will be as ubiquitous as mobile phones are today in a few decades or even sooner."* Google is heavily investing in advanced robotics and political parties are increasingly putting robots on their agenda as well. Harvard Business Review and the World Economic Forum have requested urgent research into how the displaced worker will acquire the necessary skills for new jobs.

Industrial robots have been at home in automotive plants since the early 1960s, but in the past few years the invention of more sophisticated sensors have led the robots being increasingly used in other fields like—healthcare, military and safety. Year 2017 saw a total of 381,000 units of industrial robots sold and it is expected to reach 630,000 units by 2021, according to 'World Robotics', IFR World Robotics 2018, a report from the International Federation of Robotics.[24]

Advanced robots of the digital age are performing complex and crucial roles. Robots are being adopted for fabrication and logistics to perform progressively challenging duties as they make their way into the office space. Advanced robotics is one of the top most disruptive technologies in business today. Robots with advanced senses, physical capabilities and intelligence can efficiently perform complex jobs—once thought too delicate/uneconomical to automate at optimal costs. These technologies are playing a major role in advancing healthcare. 'Robotics Surgery' is making surgeries and medical procedures comparably less invasive and more accurate, while 'Robotic Prosthetics' is restoring functions in people with disabilities and the elderly like never before. The United States, China, Japan, UK, Taiwan, Singapore, Korea, Germany, SAR, France, Israel, Netherlands, India and Spain are the leading hubs for tech innovation.

Rapid adoption of robotics by a large number of companies in recent times has washed up a new business term—Robotic Process Automation (RPA). RPA as defined by IRPAAI (Institute for Robotic Process Automation and Artificial Intelligence) is the application of technology that allows the configuration of a computer software or a 'robot' to capture and interpret existing applications for processing a transaction, manipulating data, triggering responses, and communicating with other digital systems. The manufacturing industry has been and is capitalizing on robots to increase production rates and improve quality for a long time. This new wave (RPA) is now reinventing business administration and operational processes in numerous industries. RPA dramatically improves accuracy and processing time, thus drastically increasing productivity.

Consider the fact that a software robot costs a business around one-ninth of a full-time employee working in the UK or the US, and one-third of an employee working in offshore locations like India and Philippines. This leads us to believe that RPA is not only a potential competitor but a far superior technology to business process outsourcing. RPA is set to impact the global USD 300 billion worth BPO industry. The BPO market size was valued at USD 195.2 billion in 2017, and it is expected to reach USD 343.2 billion by 2025 [25]. It demands for the reskilling of over 3 million people employed in BPO work in India and another million in the Philippines. RPA will possibly have an effect on every individual's job. Leaders in BPO like Wipro, TCS and Infosys have started reshaping their BPO business model which is focused on employing an increasing number of people to automate.

Adopting RPA as the core capability can help the industry leaders to continue dominating their industries. However, many of these leaders are not certain about the future and are risk averse, wanting to follow the tested and trusted FTE-based models. These leaders should understand that RPA is not a risk to their existing business but a huge value-add that can create growth and scalable impact.

Large well-known companies around the world are already adopting robotics into their business models. Some of these companies have found that robots are better at doing certain jobs than their human counterparts. History was created by SpaceX when its Dragon capsule docked on the International Space Station. But the robotic triumph that was part of the ordeal went largely unnoticed—Dragon is an autonomous vehicle and the company wants to keep that way, even if it is approved to carry passengers to and from the station.

Robotic process automation software created by Blue Prism, UiPath, Automation Anywhere and others, creates a digital workforce which follows rule-based business processes and interacts with the systems like everyday users. Blue Prism is one among the most successful companies developing digital workforce models. The UK-based company has expanded its operations by setting up centers across the world.

Defense contractors like Lockheed Martin have also used robotic technology to enhance their product, such as flying robot pilots for flying through unfriendly environments. Lockheed Martin along with Kaman Corporation introduced the unmanned K-Max chopper, which can deliver more than 3500 pounds of food and other items. It was deployed in 2011 to deliver supplies to US Marines in Afghanistan.[26]

At present, around 70% of industrial robots work in the automotive, metal and machinery and electrical/electronic industry sectors. However, the trends in robotic industries are changing. 'The Robot Report's global database' from an analysis of 752 robotics-related startup companies shows that, only 25% of the startups are focused on Industrial robotics while 75% of them are innovating in new areas of robotics. 25% of these are innovating in unmanned aerial, land and underwater robots for security, surveying, filming, delivery, marketing, military operations, oil and gas industry. 6% of these are innovating in agriculture, 7% in mobile robots as platforms, 3% in personal service bots, 7% in professional service bots, 7% in medical, surgical and rehabilitation, 7% in consumer products such as home cleaning and security, 9% on entertainment and 5% focus on the educational and hobby market.[27]

A study published by ZEW in partnership with the University of Utrecht confirms the positive impact on robotics job creation. The automation programs with robotics in the US had a positive impact on employment creation and the same trend has been observed in the automotive sector in Germany. Robotic and Automation market in German reached over USD 16.5 billion in 2018, which represents 4% growth.[28] *"The new record shows that worldwide demand for robotics and automation technologies from Germany remains high. However, the general political uncertainty and cooling of the global economy led to a more moderate growth than originally expected."* —Wilfried Eberhardt, Chairman of the VDMA Robotics + Automation Association.

Robots are changing the way we live, and the way businesses work in today's environment in obvious ways. Robots can do a lot of tasks better than humans can, but is a robot more intelligent than a human counterpart? This is where artificial intelligence clocks in. Like the term 'robot' itself, artificial intelligence is hard to unequivocally define, but here is a concise, comprehensive and simple definition by How Stuff Works: *"Ultimate AI would be the recreation of human thought processes—a manmade machine with our intellectual abilities. This would include the ability to use language and the ability to formulate original ideas."* Roboticists have made a lot of progress with limited AI and are approaching the level of true AI. *"Everything invented in the past 150 years will be reinvented using AI within the next 15 years"*—Randy Dean, former chief business officer, Sentient Technologies. Randy Dean utilized AI's power at his San Francisco-based company on sectors like healthcare, retail, food and financial trading market in multiple ways. In addition, Sentient, joint ventures in developing algorithms that are created by humans and are optimized for business by machines. They train AI using their historical market data to find signals that might otherwise take thousands of years for humans to find. The readily available high connectivity and secured cloud platforms are complimenting AI's growth, and its large-scale processing power is one of the main reasons it is becoming even more relevant to today's businesses.

The complicated execution of an AI problem is based on a simple problem-solving concept. AI systems gather data from the environment through multiple modes like sensors and manual input. Before examining the gathered data, it compares with the previously stored information and suggests the ideal course of action based on the comparison. The learning ability and capacity of robots are enhancing at a rapid pace. Some robots can interact socially—like Sophia, a humanoid robot created by Hong Kong based Hanson Robotics Company. Sophia is designed to be a sensible companion for the elderly at nursing homes. It aids crowds at large events or in public spaces. She has enviable human interaction capabilities and became the first robot that received citizenship when she officially became a citizen of Saudi Arabia in October 2017.

Advanced AI systems are now mastering self-improvement and learning capabilities of humans, and with these capabilities honing the ability to respond to specific situations. AI platforms are rapidly automating business processes, from capturing client information to serving as the consumer interaction interface. Although there are concerns about AI potentially causing job and employment declines, still AI should be used to automate mundane tasks with the aim to save resources for higher value work. The latest technological advancements like machine learning, deep learning and cognitive computing are complicated and expensive for many companies, but the early advances of AI can fit the budget of most 'small to mid' size companies. AI can produce great results with very few technical modifications, in less time and reduced transformational cost. Those who take these initial steps can lead in the future, as these will be prerequisites for everything that follows in the AI and digital age.

New age companies like Netflix are increasingly adopting machine learning at their core to provide a more personal experience to customers. These types of algorithms are continually gathering consumer data to give improved suggestions and help users make the most of their subscriptions. Using information from diverse datasets is a common use of AI. Under Armour is using IBM's crowd wisdom platform 'IBM Watson' to provide personalized training and lifestyle advice. Crowdsourcing will potentially become an even bigger industry as further advances in AI lead to the need for human guiding hands to adjust datasets.

Amelia, a new artificial intelligence system, was introduced by SEB (Skandinaviska Enskilda Banken) in 2016 to become a part of their consumer support team. It was tested through 4,000 conversations with 700 employees during a three-week period, acting as a part of the bank's internal IT support team. *"Amelia solved the majority of issues without delay. Customer service is a key differentiator in the competition for customers. Amelia will be an additional way for us to increase accessibility for our customers and make our service even more individualized."*—Rasmus Järborg, chief strategy officer, SEB. AI is also taking over the website design industry by storm, capitalizing on image recognition and typography selection in order to automate the website designing process.

The artificial intelligence market is expected to reach USD 190.61 billion by 2025 from USD 21.46 billion in 2018, at a CAGR of 36.62% during the forecast period, according to MarketsandMarkets.[29] The market growth is mainly driven by factors such as the adoption of cloud-based applications and services, growing big data, and increasing demand for intelligent virtual assistants. Artificial Intelligence is a promising sector for companies to invest and has immense market growth potential. According to Accenture PLC's global study, more than 1,000 global companies are already using and testing AI machine learning systems.[30] The emergence of AI is not necessarily replacing humans but is building completely new categories that require skills and training without having precedents. Machine Learning and AI were adopted early by leading high-tech, telecom, and financial services companies. Netflix has achieved impressive results by utilizing AI, enabling better search results, and helping customers quickly find their desired content. Netflix estimates that it has been managing to avoid cancellations of subscriptions, which would decrease their revenue by USD 1 billion per annum by the adoption of AI.

An integral part of AI is cognitive computing, used by AI functions. It incorporates self-learning systems which use data mining, natural language processing and pattern recognition to imitate how the human brain functions. Cognitive computing can create automated IT systems that can solve problems without requiring human assistance.

There have been great opportunities in adopting artificial intelligence, its effects on business and society. Chef, travel agent, fitness trainer, investment advisor, personal assistant and personal shopper roles have been replicated by the first generation of cognitive computing. With the potential to impact and assist humans in their professional and personal lives, AI's market opportunity seems limitless. Its potential could equal the one that was seen at the onset of the commercial web in the mid-1990s.

"Cognitive systems have the potential to radically redefine everyday life, changing how companies deliver products and services, engage and interact with customers, learn and make decisions,"—Forbes. Governments and many businesses are using these technologies, including automotive healthcare, tourism, media, retail, transportation, engineering, law and pharmaceutical organizations.

Continuous availability of data on business processes is essential in this data-driven era of cognitive computing. Through the advanced analytics and automation, it is possible to predict potential issues. When cognitive computing systems learn a domain, they build up additional knowledge over time and continuously get better and smarter. Their learning process primarily includes language, parlance, processes and interacting methods. Cognitive computing systems have the ability to process natural language and unconstructed data, unlike systems of the past that required rules to be hard coded by a human expert.

IBM's tech capabilities help clients transform from reactive to predictive business model, which aims to predict and avoid disaster before it occurs. IBM started its Watson research project to build a computer system that can learn on its own and can think like a human. Watson competed in and won a high-profile 'Jeopardy' contest in 2011. It combined natural language processing, machine learning and knowledge representation in a way that no other systems had in the past. Watson processed the questions, scanned its database for information, developed hypotheses, analyzed the potential outcomes and produced answers that were also in natural language form. IBM implemented Watson for many cross-industry domains like financial services and healthcare over the past few years. The impact of Watson has been huge in the business world. IBM joined with the Government of the United States to bring Watson technology into America's healthcare network and its largest hospitals. The US Department of Veterans Affairs Hospital is installing Watson to help soldiers get the best-suited cancer medicine by providing genomic analyses through cross-referencing DNA sequencing data.

Many cognitive computing startups are creating a sales pipeline for companies through industry-specific apps. The industries that generate high volumes of unstructured and structured data are best suited to benefit from cognitive computing. These include retail, healthcare, energy and financial services. Companies like Accenture and Deloitte are investing more and more resources in cognitive computing and are setting up client innovation labs to innovate and develop cognitive systems for clients in no time.

According to CB Insights, there has been about 7000+ AI Acquisition deals across 20 industries and 12 cross industries application since 2011. Over 3600+ AI startups have raised equity funding globally. The AIs are playing the role of 'Software-As-A-Service' in business process automation (UiPath, Blue Prism, Automation Anywhere, Face++). The AIs are functional in developing core products like leveraging machine learning to develop microbial seed treatment (Indigo Agriculture); The AIs are being implemental in hardware to support AI workforce (Graphcore, Habana, and Cerebras).

For instance, under core industry like Finance & Insurance, startups work on applications including compliance, fraud detection, back office automation, personalized finance management, underwriting, and chatbots specific to the finance and insurance industries. Under cross-functional, examples include companies developing applications like computer vision, industry-agnostic speech analysis, natural language processing (and generation), language translation, transcription services, facial recognition, object recognition, and conversational AI platforms.

Tech giants like Facebook, Amazon, Microsoft, Google, & Apple (FAMGA) have all been aggressively acquiring AI startups in the last decade. 635 AI acquisitions have taken place since 2010, as businesses aim to build up their AI capabilities and attract the best talent. Apple made a phenomenal 20 acquisitions and stands at Top 1, followed by Google (14 acquisitions), Microsoft (10 acquisitions) and so on.

Some of the top trends in AI are open source frameworks, facial recognition, edge computing, medicinal imaging and diagnostics, predictive maintenance, e-commerce search, crop monitoring, autonomous navigation, reinforcement learning, network optimization, cyber threat hunting and language translation.

7

Data Transformation, Intimacy and Migration

"Leaders base decisions on Data, not Intuition."
— **Avinash Vashistha**

"Big data" is one of the biggest challenges and an equally big opportunity that enterprises face today. According to Forbes, total digital universe of data has grown 10 times since 2015; there are over 6 billion mobile phone users and still less than 0.5% of all data is ever analyzed and used! The volume, complexity and variety of data available is growing at an astounding pace, which helps explain that last startling statistic: enterprises gain insights from less than 0.5% of this data.

In today's world, companies need to master the art and science of data transformation and management. Data transformation, is an ETL (Extract/Transform/Load) process, involving converting a raw data source into a sanitized, validated, and ready-to-use format. It can turn data into timely insights that positively impact businesses and provides the much-needed competitive edge. Data should be accessible, consistent, secure, and be trusted by the users, auditors and relevant regulatory authorities.

Data transformation requires a conscientious data strategy that will deliver business value to all its stakeholders. According to Import.io's paper, "Top 7 Best Practices for Data Transformation" published May 10th, 2018, key best practices evolved from experience across a diverse portfolio of enterprises are:

1. **Start with the end in mind – Design the target:** Enterprises need to engage business users to understand the business processes that they are trying to analyze, and design the target format, before data transformation can deliver insights. The above "Dimensional Modeling" process, needs to deliver "Dimension Tables", which provide the "Who, what, where, when, why and how" context for the data. The second element the above process delivers is "Fact Tables", which store the results of the events being measured, and answer the "How many" questions.

2. **Speed date your data with data profiling:** Knowing the business process you would like to examine typically leads to the source(s) of data to be turned. To evaluate market patterns, for example, you would need to enter the customer database, the inventory database, and then extract sales reports from a point-of-sale network. Once the root of the data is known, the raw data can then be converted to a usable format.

3. **Cleanse – When your data needs a bath:** Equipped with data modeling tools, you can better understand how much and what kind of data engineering work you need to do on the data in order to make it useable. For example, if the date fields of the source data are in the format YYYY / MM / DD, and the target date fields are in the format MM-DD-YYYYY, you will have to convert the source date fields to suit the target format.

4. **Conform data to the target format:** The above three phases are the inceptions for data transformation into target format. Here the understanding of the source data from the data management department satisfies the need for data attributes for the consumers to evaluate a business process. Starting with mapping source columns to target columns, the data transformation team makes use of ETL tools to automate the data flow on successive data loads for those columns.

5. **Build dimensions then facts:** Dimensions place meaning around the data; the details explain what happened within the dimensional context. For example, customers, products, and dates could be dimensions; sales results and measures could

be facts. Secondly, the benefit of loading dimensional tables is that, freshly loaded fact records can then connect to relevant dimensional records. Sales data would not be very helpful if there were no ties to measurements of consumer, commodity, and time. Therefore, consumer, product and date dimensions must be updated with each data load first, preceded by the selling information sheet.

6. **Record audit and data quality events:** During the cycle of data transition, audit tracking and data quality measures have huge benefits. Audit tracking measures the number of records enabled at each step of the process of transition and the period such changes took place. Capturing data quality test results, including in the data load audit documents, and connecting fact records to audit records provide the opportunity to trace the history of fact data and help to prove the validity of measurements measured from fact data. This approach allows analysts to "work backward" to react to specific stakeholder questions such as "Where did this data come from?" and "How do I know that those metrics are correct?". Getting ready and accurate answers to these questions creates customer trust in the transformed data and offers solid ground for continued interaction with end users with the data transformation team.

7. **Continually engage the user community:** The true indicator of the importance of data transformation is to what degree the intended user community embraces the converted digital commodity and consistently utilizes it. But rendering freshly minted, conformed data available to end-users isn't the end of your data transformation; it's just the end of the process. Transformed data must be thoroughly tested for customer approval, and the data management team must quickly address faults discovered by business users "in the wild." The abundance of data today is a potential gold mine for businesses. And, like gold, to optimize its worth, such data must be extracted analyzed, refined and delivered to maximize its value. Understanding the fundamentals of data transformation, such as dimensional modeling, profiling cleaning conforming, testing and presentation, can allow you

to discover valuable insights from your data that can affect your company greatly.

"Data Intimacy" Culture

Data-centric leading companies need to maintain a tradition of "information trust" within their organizations. This requires a change in mindsets, attitudes and behaviors. In a Data Culture, people ask hard questions and challenge ideas. Through behavior, the leaders encourage judgments based on data, not instincts. Which needs leadership to create a "Digital Intimacy." According to Tableau, organization's five main tenants are – Trust, Commitment, Knowledge, Communication and Mindset.

1. **Trust – Leaders create a foundation of trust in people and data:** Trust is the foundation of a strong Data Culture. Leaders are trusting their people; people are trusting the data and trust each other. Right model of data governance creates a single source of truth that breaks down silos around departments to build high-confidence, collaborative ties. The company provides data insights to discover impactful approaches.

2. **Commitment – People treat data as a strategic asset:** Organizations with effective Data Cultures, devote themselves entirely to understanding the importance of their information assets—and help people make better decisions through data. This dedication is apparent in all facets of the organization—from the framework of the company to daily operations.

3. **Talent – Organizations prioritize data skills in recruiting, developing, and retaining talent:** Ultimately, Data Culture is made up of data people. Even with the best technology, system and processes, they can't be powered by data, if people don't understand how to deal with the data. As part of the talent strategy, managers must emphasize data skills for recruitment and training—clearly outlined in job descriptions and defined in the hiring process. Everyone in the company must feel confident that they will find the right

results, apply analytical methods to their job and communicate their findings.

4. **Sharing:** Most data-solving challenges aren't limited to a single unit or business line. We need data from multiple platforms and various teams working together. People have a common interest in a data culture— using data to improve the enterprise. Together, people amplify the impact that data can have. The cooperative attitude builds "stewardship" around the data and analytics, and cultures.

5. **Mindset:** The development of a data-first mentality is equally important as the development of data skills. In a Data Culture, data is given priority over intuition, anecdotes or rank. Data mindset exchanged by all, creates open discussions and ideas that contribute to exploration and innovation. In such an environment, data is seen as a source of personal growth and career development. People are curious and willing to challenge evidence with their own hypotheses— and are open to challenge from others. When data-driven practices become habits perceptions change and people begin to equate data with performance and development.

The Art of Data Migration (SAP Legacy to S/4HANA)

Data migration between Enterprise Resource Systems (ERP) is a critical business reality. One of the major ERP migration events happens in 2027! SAP will stop supporting its Business Suite ERP system, with users needing to upgrade to S/4HANA instead, in 2027. Two-thirds of SAP customers are planning their company's future with SAP S/4HANA.[31]

There are compelling business drivers for Data Migration. The shift to SAP S/4HANA is a perfect example. Progressive organization understand that making the switch now will save a lot of pain further down the line when SAP stops supporting older systems. Yet it is not just about the availability of the S/4HANA upgrade. It is actually about introducing more efficiency and productivity into operational workflows as soon as possible.

However, these business gains are at risk if the data migration is not well thought through. Poor data migration planning and execution leads to:

- Potential 'Black Swans' – the chance that something big will go wrong.
- Project overrun – extra time that diverts resources from other projects.
- Additional budget – mounting costs that may lead to a decision to quit.
- Lost ROI – lengthy time-to-value reducing the intended returns of S/4HANA.

Any new system is only ever as good as the data you put in. So how can you plan and execute a data migration without these risks? According to Syniti, there are eight key steps to a successful data migration:

1. **Prepare – Define your team roles, systems and schedules:** This first phase is the building block of future success. It points out how you will manage the data transfer with the wider business goal in mind. Progress indicators include: setting up the reusability initiative; identifying potential economies of scale; giving clear roles and responsibilities across the project team; outlining the special requirements (e.g. audit reporting); and focusing on staff, systems and technology.

2. **Extract – Collect and 'stage' your source data:** This is often taken for granted stage but rarely executed to the full. It can reduce the time and resources needed for further phases to get it right. Progress indicators include: visibility of all data held by the organization; all data sources defined prior to the Profile point, and awareness of the meanings of the underlying data source.

3. **Profile – Analyze and cleanse your data:** Profiling provides insight into the legacy data and identifies the essential fields required for S/4HANA migration. Progress factors include: only business-relevant data is migrated; the

method helps to understand the pros and cons of current workflows; and results provide for clean-up deployment of the right resources.

4. **Design – Create your S/4 HANA blueprint:** The configuration of the target data would reflect the new system whatever the source. Doing so holds the migration on track even as S/4HANA's architecture keeps changing. Progress factors include: architecture determines the optimum data that S/4HANA requires; regulations can be replicated during go-live to save potential resources; and it is a method of coordination between existing and future data.

5. **Map – Align your sources to S/4HANA fields:** The stage may sound straightforward. Yet many data projects fail because the mapping process is performed in hundreds of spreadsheets which cannot be easily managed. Success factors include: consistent design rationale between the legacy sources and S/4HANA; quick testing of mapping completeness; reusable processes and rules are specified within the mapping.

6. **Construct – Manually enrich your data:** Most datasets will need to be reinforced with business user information. SEP For those non-technical colleagues, this stage should be made simple to maintain progress. Success factors include: data construction is a carefully controlled process; business users are empowered to look after their data; data can be found and corrected from one location; and business users can find data quickly and re-mediate.

7. **Transform – Simulate and validate your loading process:** The Transform stage provides the quality data required by businesses in the future. This combines all the results of the previous phases together, so data is primed, instead of a go-live. Success factors include: validating data to ensure that it can be enabled, reconciled and used; implementing correct rules and settings for source and goal data; and eventually verifying the quality of the data by testing.

8. **Load – Move your data into S/4HANA:** Once all business and technical reconciliations have been done, the data can be trustfully transferred to S/4HANA. Success indicators

include: high quality data, user acceptance of data and few or no business interruptions. You will convince the clients by pursuing each of these steps that your data transformation has been well planned and well performed. You're going to have peace of mind that you've fully planned for your transition to SAP S/4HANA. As a consequence, the company will enjoy the benefits before those who postpone their transfer or fail to follow best practice.

A successful data migration platform - example, Advanced Data Migration (ADM) from Syniti, solves complex enterprise data transformation challenges with all eight stages of a successful migration built-in. Leading migration platform and solutions combine Intelligent Automation and Machine Learning with expert insights gained from thousands of data migrations. They provide project oversight, visibility; knowledge preservation for future migrations; speed and agility in implementation. A successful data migration engagement augmented by Intelligent Automation, should significantly automate the process >80%, save clients up to 50%, increase quality and provide a ROI in less than 9 months.

8

IT Infrastructure and Business Process Automation

Author: Venkat Thiruvengadam
Founder and CEO, DuploCloud Inc.

"If you want something new, you have to stop doing something old."
— **Peter F. Drucker**

Digital transformation is about changing the business processes in an organization to adopt modern cloud solutions that result in substantially more automated and time efficient workflows leading to labor optimization and increased productivity. Companies adopting digital transformation become more nimble, are able to scale rapidly, grow their business and fend off competition from non-traditional rivals who are natively digital.

This business transformation is not possible without fundamentally rearchitecting the underlying technology infrastructure that drives the business processes.

Business process automation is typically viewed in the context of specific vertical and horizontal industries. There have been several software innovations like SAP, RPA, solutions in supply chain, healthcare mobility solutions, financial services (Fintech) etc. Infrastructure modernization is often ignored or punted to the consumer. Without a native and out-of-box supporting solution to IT, many innovative and specialized software applications aimed at automating business processes take too long to deliver, thereby diluting the return on investment.

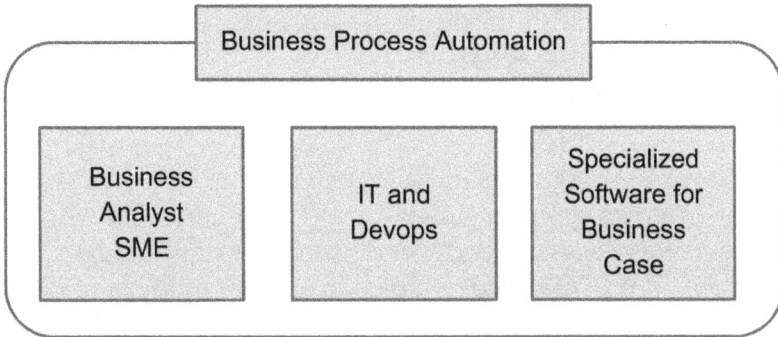

```
┌─────────────────────────────────────────┐
│         Business Process Automation      │
└─────────────────────────────────────────┘
 ┌──────────────┐ ┌──────────────┐ ┌──────────────┐
 │   Business   │ │   IT and     │ │  Specialized │
 │   Analyst    │ │   Devops     │ │ Software for │
 │   SME        │ │              │ │  Business    │
 │              │ │              │ │  Case        │
 └──────────────┘ └──────────────┘ └──────────────┘
```

Case Study 1:

Success of managed services by Amazon Web Services (AWS)

AWS has many non-IAAS products (beyond Storage, Compute and Networking) like databases, IOT platform, AI/ML solutions, Analytics etc. All of them are "managed" i.e. developers can directly consume them without having to worry about the underlying infrastructure. The claim here is that, AWS solutions for vertical industries are successful, not just because the subject matter content is great but also because they are "managed". In fact, much of public cloud's success can be attributed to the infrastructure problem they have solved at various business layers.

A relevant example is of Confluent, a multi-billion-dollar valuation company which, at its core, is adding an infrastructure management layer in top of open source Kafka. In fact, many of the successful functional software have existed prior to the cloud and the only value that the public cloud adds, in this context, is that it takes care of the infrastructure which, in itself, is a massive value proposition because of the large business scaling possibilities.

Case Study 2:
Failure of Independent Software Vendors (ISVs) in building their next generation cloud offering

Last year we were commissioned, by two distinct ISV (Independent Software Vendor) company CTOs, to do a postmortem of why their effort to build the next generation "cloud" version of their well-established on-premise product failed in spite of millions of dollars in investment. Infrastructure issues were at the core of both the failures. Most ISVs effort to build the next generation "cloud" version of their well-established on-premise product fail, in spite of millions of dollars in investments. Infrastructure issues are at the core of majority of such failures.

They got trapped into a very common pattern we are seeing increasingly:

1. The goal was to build a cloud ready product. Any literature on the internet talked about Microservices, Docker, Lambda, Devops along with Multi-Tenant SAAS.
2. To be fair I think the ISV Companies knew their core business was single tenant and not many clients would buy a hosted multi-tenant SAAS. "But will this solution work for the core audience?" This question has not been properly addressed by the ISV industry.
3. Fast forward 2 years, many millions spent, Multi-Tenant SAAS is not selling. It turns out the underlying infrastructure has 17+ micro-services, all the latest and best tools are in use, but specialized devops teams are required to run the infrastructure.
4. The net result: The "SaaS" product cannot be shipped to customer who do not have the know-how or the will to operate such an infrastructure in the cloud. The new business features that were linked to this "SaaS" release also get lost from the core enterprise solution.
5. Next gen cloud platform dies, the company decides to ship the existing on-premise solution with a reference architecture for customers to download and install from AWS and Azure

marketplace. This approach is a very painful process for customers and business growth slows down

The conclusion here is how an ill-informed infrastructure architecture with a noble intent can fail a product line.

Case Study 3:
Economic loss and delayed ROI with enterprise applications

Many modern-day applications from companies like SAP, EMC, Cisco, VMWare, Adobe AEM, Kafka, RPA software provide great value to their users, but installing and maintaining them requires specialized skills. Each one of this software require certified engineers, and several months to do the deployment, before business value can be realized. In the overall ROI analysis, the cost of operations and opportunity cost due to long implementation cycles, makes the value of such ERPs questionable.

Next generation enterprise applications will be simplistic to deploy, and/or the software will ship with cloud automation infrastructure bot, that can manage itself. Alternatively, the software vendor will have a "managed service" offering that would scale to thousands of clients.

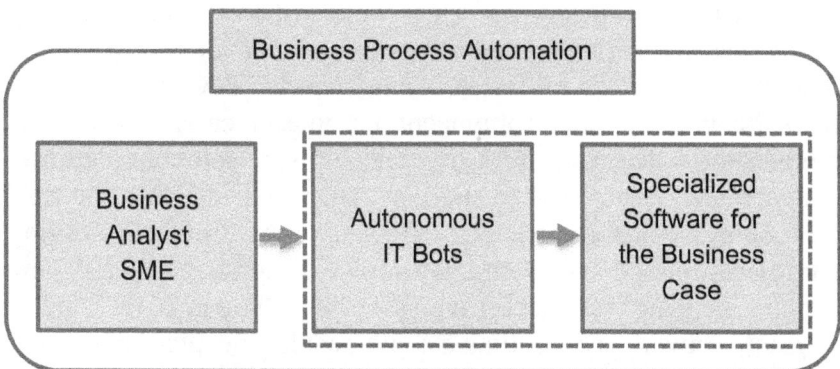

9

Digital Industrialization
- Internet of Things (IoT)

"Industrial companies are in the information business whether they want to be or not."
— Jeff Immelt

The Internet of Things or IoT in short, is the trendy topic of conversation, both on and off the workplace. *"It's a concept that not only has the potential to impact how we live but also how we work. This is the concept of basically connecting any device with an on and off switch to the internet or each other."*—Forbes.

Everything from our phones, coffee makers, washing machines, televisions, wearable devices, almost anything one can think of, can be connected to the internet. Machine components like a tool holder of a welding machine, the drill of an oil rig can also be included in the category of being connected to the internet. Any device that has a switch/controller can be a part of IoT. By 2025, there will be over 75 billion internet-connected devices.[32] The IoT market constitutes node component, net in each case, new insights are generated that work infrastructure, software solution, platform, and service. According to Fortune Business Insights, the IoT technology market is expected to grow at a CAGR of 24.7% during the forecast period and reach USD 1102.6 billion In each case, new insights are generated that 6 from USD 190 billion in 2018.[33] IoT is a giant network connecting people-people, people-things, and things-things. As the cost of connecting is reducing due to cheap and widely available internet, many devices are getting created with built-in sensors and Wi-Fi capability and is a great opportunity enabler for the IoT.

More connected devices mean the businesses get greater intelligence about their operations and hence the ability to improve their efficiency and effectiveness. *"An intelligent system is transformative. For an efficient inventory control mechanism point-of-sale scanners are connected to warehouse systems and analytics software at headquarters for industry. Robots on a factory floor send production and maintenance information directly to those who need it, for unparalleled reliability and uptime. In each case, new insights are generated that drive the organization's objectives forward on many levels."*—Microsoft report on IoT.

Cybersecurity is the drawback of this excessively connected world. The opponents' threat of hacking, by stealing valuable and sensitive information is greater, as the data collected online is more. An article on TechRadar focused on how small and mid-sized businesses will have to consider securing their data collection methods, the article states—*"Connected devices are considered vulnerable to hacking because many of them use the Linux operating system which isn't patched in the same way as licensed services such as Windows. Therefore, firms must approach IoT in the same way they do for all IT systems, by updating them regularly and using identity management and authentication."*

Even with the security risks, the IoT will allow companies to make smarter products. Earlier phones were only used to make phone calls; user expectations from these devices increased. Having a smart tennis racket, smart yoga mat, or internet enabled frying pan, might seem strange but these are the forays into the world of IoT.

It is not only about designing smarter products, but companies can also engage in smarter business operations and make smarter decisions. Everything from yogurt cups to the cement in bridges can be attached with IoT sensors to record and send data back into the cloud. Collection of specific feedback from this can help one understand how products or equipment are used and when they break. They also provide insights about what people demand for future. For example, in Rolls Royce aircraft engines, the IoT sensors sends real-time data on functionality of engine to the monitoring ground stations, which can be used to detect any fault before it develops a catastrophic failure. To eliminate the least popular features from its products and focus on enhancing more popular features, Microsoft uses software that constantly collects data on what features are being used the least.

For some business, possibility of a change in business model can also be sensed by IoT. Take John Deere; for decades, they have sold the tractors that have made farming scale-up with more profit. In 2012, the farmers were given information about the best suited time and method to plow, and best route to take while plowing by addition of data connectivity to their devices. They are also into the business of selling data as much as they are into the business of selling tractors.

It appears to be a crucial change in the manner we see the world. There comes a day where we won't be able to imagine a world without smart cars, smart roads, smart infrastructure, in the same way where we can't imagine using a phone which is just a phone. The new rule for the future is going to be 'anything that can be connected will be connected.'[34]

Conversations about how IoT is going to impact all our lives, is taking place all over the world. Analysts are also trying to understand what may be the opportunities and challenges that would occur when many devices are connected through IoT. As of now, educating ourselves about IoT and its possible impacts on how we work and live is the best thing we can do. In other words—The Internet of Things is changing everything around personal lives and businesses; every business needs to align its effort to benefit from IoT.

The Internet of Things is attaining importance in many industries. As an example—monitoring the energy grid and other supply networks for efficiency and to troubleshoot problems that could shut down operations. Manufacturers are using IoT to create networks that monitor processes for safety and efficiency and help refining process control. In medical and healthcare industry, IoT is enabling remote patient monitoring and emergency care. It can also be utilized in planning the construction of commercial and residential buildings, safety at site, automating processes like inventory management, creating modules that are manufactured in a plant and then assembled into a complete structure at the location by the builders.

Mining companies are using IoT capabilities to connect the latest technology advances and development of virtually everything associated with the operation. Products like vehicles, heavy mining equipment, and even the sonar used to assess the potential of a drilling site, which were considered as standalone elements, can all integrate and work together using the IoT. This seamless experience means more data and less guesswork. The technology allowing driverless truck in the mining process is employed by companies like Rio Tinto. The same company is experimenting with Autonomous Drilling Systems that can operate without direct intervention by humans. BHP Billiton is also running driverless trucks at some sites.

IDG in a partnership with Intel surveyed 200 IT leaders on the subject of IoT. The outcome of the survey was that the IoT is becoming the area of interest to IT leaders, with the majority of respondents defining it as the 'the biggest revolution in business computing for decades.' However, what the survey also found was that there is a perception that only a handful of sectors will benefit from IoT technologies. Respondents felt that the top four industry sectors that will gain from the Internet of Things, in terms of efficiency and innovation are the technology, consumer electronics, energy and automotive.

One such example is the number of successful IoT-equipped Smart Buildings that are being built. Intel, Tatung and Elitegroup Computer Systems are working together on such a project, focused on creating smart technology which will enable them to save energy for businesses. The IoT solution uses an IoT gateway, integrated with an I/O board that provides a smart connection between energy management systems and other commercial building devices. The initial verification phase produced 'excellent' performance from a smart, energy-saving conference room. The technology is addressing an entire smart building, targeting a 30% improvement in building energy efficiency.

In manufacturing, Intel and Mitsubishi Electric worked together to create advanced factory automation systems that use end-to-end IoT connectivity and big data. Mitsubishi's Intel Atom-based gateway enables secure gathering of aggregate data for the analytics.

Data is then processed using software from Revolution Analytics, hosted on Cloudera Enterprise, the foundation of an enterprise data hub. The pilot resulted in improved equipment, and increased yield and productivity. In addition, Intel gained the ability to conduct predictive maintenance, with reduced component failure, and optimized inventory of tooling and human resources. With the help of this IoT based solution, Intel realized USD 9 million in savings through cost avoidance and improved decision making.

There are many examples of industrial uses of IoT technology, but agriculture is often one not immediately thought of. Abbaco Controls in Malaysia is using IoT technology to create efficiencies through the remote control of water levels in its paddy fields. The solution features real-time sensor networking and data transmission, which provides accurate and real time status of water levels, and water flow both in and out of the fields. This enables user to monitor and respond quickly and remotely via smart mobile devices. This has drastically reduced the lead time to farmer's request, minimized manual intervention, and ultimately, increased operational and farmer efficiency. Data analytics and reporting at the back end has led to further efficiencies. Abbaco controlled IoT system is the first water irrigation automation project in Malaysia's paddy fields.

In retail, coffee company Costa uses connected retail kiosks to improve profitability at its self-serve espresso bars. Its innovative coffee kiosk, the Costa Express CEM-200, uses bright and colorful high-definition touch screen displays, cashless payment, telemetry, near field communications and digital gestures signage. Behind the scenes, the station utilizes anonymous viewer analytics—part of the Intel audience impression metrics suite, that gives retailers a better understanding of their customers and their purchases. This enables them to optimize their product mix by location and customer type, charge premium prices and up/cross sell, and share information with the buyer. Costa found that the stations have the ability to increase their sales per transaction because the touchscreen interface and cashless payment make it easier for customers to buy more items thereby boosting sales.

Even wearables are using IoT to improve its service. Of all the IoT startups, wearable maker Jawbone is probably one of the largest funded entity. It stands at more than one and half billion dollars. Wearable technology is a hallmark of IoT and the most ubiquitous of its implementations to date. The efficiency of data processing achieved by various wearables such as smart wrist wear, hearables and smart glasses is removing the inert skepticism among the public and is bringing exceptional value to our lives.

Industrial internet, connected homes, transportation, connected cities, healthcare, wearable technology, and oil and gas are the early adopters of IoT technology. At present, connected homes is the biggest segment in the market of Internet of Things. The connected homes market size—Lighting Control, Security and Access Control, HVAC (Heating, Ventilation and Air Condition), Entertainment, Home Healthcare, Smart Kitchen and Home Appliances are expected to grow from USD 76.6 billion in 2018 to USD 151.4 billion by 2024, at a CAGR of 12.02%.[35] From the jobs report in December 2014, IoT created 252,000 US Jobs in the same month, and the unemployment rate dropped to 5.6%, this shows that IoT increases job opportunity both in private and public sector. IoT job creation will contribute an additional 14 trillion to the gross domestic products of 20 largest economies of the world by 2030.[36] IoT will also open the doors for more corporate training sessions, creating an additional set of jobs–IoT trainers, IT personnel, IoT specific engineers and the creation of entire vendors which are focused on improving IoT. This IoT wave is an open opportunity for semiconductor companies who can add innovation to original device manufacturing companies and others that are building Internet of Things products and applications. There is no doubt that IoT is changing the way business is done. But, by far one of the biggest impacts it has had, is in the transportation industry. According to Andreas Mai, Director of Smart Connected Vehicles for Cisco Systems, The Internet of Things (IoT) is set to create USD 700 billion in benefit for personal transportation alone. The benefit to the tune of USD 1400 per connected car a year. That breaks down into lower costs for drivers through lower insurance rates, lower operational cost and less time stuck in traffic which increases productivity. The benefits to society are increased safety–80% of crashes can be affected by vehicle-to-vehicle communication alone. There is reduction of CO_2 and decreased congestion as connectivity helps increase lane capacity and intersection throughput, lower infrastructure costs from knowing when and where to send trucks out for ice prevention, fixing potholes, and more. [37]

The connected car has also spawned a whole new set of service providers to create helpful services through connectivity–like parking spot locator services, more accurate, real-time traffic information, and location-based services that create value. In the coming days urban transportation will become autonomous, electric and on-demand. This is the future of mobility and companies like Tesla, Uber and Google are already working on it.

Navigant Research believes that on-demand shared transportation services will merge with the vehicle technological trends of electrification, wireless connectivity and autonomous driving capability which creates a low carbon transportation system for cities over the next 25 years. This is already happening. The major constituent of technology for upcoming days is the wireless connectivity, which leads to transformation of personal transportation to mobility as a service. Navigant Research forecasts that by 2025, more than 1.2 billion vehicles will be globally connected to their surrounding and other vehicles.[38] Safety alerts, traffic notifications are provided by IoT systems, it also supports semi-autonomous driving features in advanced systems.

Electric vehicles are an increasingly popular choice in car-sharing plans in cities. Usages of electric transports are supported by city officials to control pollution in congested cities. For instance, officials in London pushed hard to bring schemes similar to the successful Autolib, an electric car-sharing service inaugurated in Paris.

Combining IoT with blockchain or BIoT—ushers in a whole host of new services and businesses. For example, BIoT can be used to track shipments of pharmaceuticals and to create smart cities in which connected heating systems better controls energy use and connected traffic lights better manage rush hour.[39]

IoT encompasses many emerging technologies like connected cars, wearables, drones and robotics. Increasingly, connected sensors are being applied to heavy machinery, supply chains, and factories leading to new digitization called Industry 4.0. IoT is allowing industrial players to derive useful insights, make decisions, and optimize their systems by augmenting with worker wearables and collaborative robotics.

Robotic automation, and the rise of robots in enterprises are allowing future of factories to approach the technology wave. According to CB Insights, Internet of Things startups have attracted over USD 7.4 billion in cumulative investment of over 887 deals, since 2010. From auto and healthcare, to insurance and heavy industries, IoT is weaving together connected devices and objects to transform how industries work.

The emergence of IoT had created a disruptive environment, characterized by value chains, business model transformation, and a "collaborate or die" ecosystem, a reality. IoT is a broader area which is not dependent only on one product or project. The thought of connected sensors and smart devices, which are autonomous, will continue to flourish, without depending on the fall or growth of smart appliances. The IoT depends on increased machine to machine communication based on cloud computing and sensor networks which collect data. It is mobile, virtual, instantaneous connection; and makes everything from streetlights to seaports in our lives, "smart."

10

Cybersecurity

"Hackers have already breached Internet-connected camera systems, smart TVs, and even baby monitors."
— New York Times.

Information stabbing, phishing, information breach, and computer viruses are threatening. Data exploitations are due to increase in digital applications, network and the number of mobile users. *"Cybersecurity, also referred to as information technology security, focuses on protecting computers, networks, programs and data from accidental or unauthorized access, change or destruction."*—The University of Maryland.

Hospitals, corporations, financial institutions, governments, military, and many more businesses gather process and store large confidential data on computers and send them to other computers across various networks. There is a necessity for the installation of a powerful cyber defense system to safeguard sensitive personal and business information from increasing cyberattacks.

The cyber risks are of three types. Foremost, is the cybercrime carried out by the person or an organized group to cause a disturbance–stealing data, extracting money, ruining operations of a website, acquiring intellectual property/credit/debit cards. Second, is a cyberwar–when a nation carries disruption and surveillance against another country to extract data/cause a disturbance. Finally, cyber terror–the organization plans the terrorist activities in a nation-state through cyberspace medium. The ThreatMetrix 2018 Cybercrime analysis stated—among 17 billion transactions processed included 103 million mobile attacks and 3 billion bot attacks. There is a growth of 67% in financial services transactions and increase of 13% year-on-year.[40] The hacking software with total support services can be purchased off to notify cyberattacks. In parallel to the revolution in technology, the cybercrime world never quit inventing. The never-ending list of bugs, threats, and viruses is published by Microsoft, Apple and other software providers on continuous basis. Cybercriminals use several ways to attack from a remote place, such as the use of a virus that obtains access to a targeted computer to steal information, files and corrupt data. Other malware includes worms that exploit using the weaknesses in operating systems and spyware and tries to take control of the processor by which the information can be stolen.

The attack vector is a path by which a hacker can access a computer/network. These include phishing (an attempt to acquire users' information), pharming (an attack to redirect a website's traffic to a different website), drive-by (opportunistic attacks against specific weaknesses within a system), and social engineering (exploiting weaknesses of the individual by making them click malicious links). Nowadays, considering the internet 'as a safe place' is a myth. Malware is found in all the junctions of Networks. There are high possibilities of computers being affected by a simple mail opening/navigating to a website. Tricksters and malicious hackers are creating various applications for hacking and circulating malware to make money. To obtain access to the computer, some malware with fake antivirus applications mask as a real application. For example, the pop-up generated by a fake antivirus software can state that the user system can be kept safe by purchasing their fake software products.

Various types of malware use unconventional methods to corrupt processers. For instance, Spyware and Trojan viruses affect the system upon installation. The presence of malware occurs due to downloading from unknown sources, whenever the user visits a website that is corrupted. These worms serve to be dangerous as they can access any system without any user interface. The software programmed by the hacker scans for the vulnerable system on the internet. The main indication of malware infestation is that the system becomes unstable. The system may reboot, crash or process slowly without any appropriate reason. The worm may cause the slow-down of internet connection which is the sign of a system being attacked. However, some complicated malware has indications that are almost impossible to detect. Malware can be programmed to send the text log, typed by the hacker. The text log can then be scanned for the password to gain access to bank accounts/personal emails. The only solution for the corrupted computer is the "Antivirus" software. A complete antivirus solution can detect various kinds of malware and is capable of deleting and cleaning all such malware. In a certain scenario, reinstalling the OS is the only solution to remove the malware.

More than 84 million new malware samples were detected and neutralized by PandaLabs throughout 2015. It means that there were about 230,000 new malware samples produced daily to contradict cybersecurity attacks.[41] According to 2019 Official Annual Cybercrime Report, *"Cyberattacks are the fastest growing crime and predicted to cost the world USD 6 trillion annually by 2021"*–Cybersecurity Ventures.[42] Banks, celebrities and corporations are not the only targets, but so are individual users. As long as one is connected on the internet, he can be a victim of cyberattacks at any place and any time.

In the FBI'S most wanted list under cybercrime, there are 69 cybercriminals listed who are responsible for the huge consumer losses. These criminals are from all around the world and rewards are offered for their capture. 'Jabberzeus Subjects' being the most wanted-after criminals of the FBI (Federal Investigation Bureau), is involved in the mass money laundering venture that installed malicious software "Zeus" on the host devices without authorization. This kind of financial malware was employed to gain access to online banking accounts, by obtaining a personal ID number, bank account numbers, and other personal data. Zeus and one of its trojans called GameOver Zeus, corrupted as many as 1.2 million computers and caused more than USD 100 million in damages.[43]

The most expensive computer virus of all times caused damage worth USD 38 billion and this virus has a very apt name—MyDoom. The most devastating computer virus till date. The creator of this virus was not found, though it was able to find the origin to be in Russia. This worm was primarily camouflaged as spam transmitted by email. Without any hint, it would forward to each address available when the user negligently opens the attachment in the email. The payload of the original version performed two things: it led away into the host computer, allowing it to control from a remote place, secondly, carrying DDoS attack (Distributed Denial of Service) against SCO (Shanghai Cooperation Organization) association's website.

There are over 4.39 billion internet users and about 3.48 billion social media users all around the world according to the 2019 study[44]. Social networking is commonly used to stay connected with family and friends and to spend time. This certainly attracts cyber hackers towards social media. Users on social media are more likely to check for the links that are posted by friends which serve as an advantage to hackers. Cyberattacks in the social arena include things like "Like-jacking", where criminals post fake Facebook "like" buttons to web pages that install malware when clicked on. Phishing, where there is an attempt to acquire sensitive information such as usernames, passwords and credit card details by masking itself as a trustworthy entity in a Facebook message/Tweet. Because social media users trust their circles of online friends, the result is more than 600,000 Facebook accounts being compromised every single day. The victims of cyberattacks are in ration 1:10.

Commodities such as Adobe Reader, Adobe Flash Reader, and Oracle Java are present in most of the computers. This shows that 99% of users are exposed to attacks. This software provides crucial vulnerabilities—just a click on the corrupted advertising banner transfer complete control to the hacker over the system. Adobe Flash is aimed primarily because of its extremely fragile nature during many of the attacks. The attackers can crash any system due to the security dent in Flash with several CryptoLocker alternatives or CRB Locker and Teslacrypt. The increased use of automation and abuse kits has led to heavy and advanced attacks. Even if the system and browsers are carefully shielded, an individual can be open to a range of cyberattacks.

The chances of stealing secure corporate information by the employees who are fired from the organization are about 59%. Indeed, there should be protection against insider threats. The malware presents internally, though considered being less frequent, can affect the system at their access level. The risks faced by an authoritative head with special identities are high. *"Data breaches that result from malicious attacks are most costly"*—Ponemon Institute. The insiders can be deceived by external agents to provide passwords or data. The accidental modification and deletion of crucial information may happen by careless insiders by using the wrong key. *"Almost half of the European organizations believe that insider threats are now more difficult to detect, with senior IT managers being very worried about the things their users can do with corporate data."*—Andrew Kellet, Principal Analyst at Ovum.

America witnessed considerable rift when Target Corp misplaced credit card numbers of USD 40 million worth to Russian hackers in late 2013. Adobe Systems, J.P. Morgan Chase, Anthem, Home Depot, and eBay are some companies that have been subjected to the cyberattack. One of the hacking instruments found by Hewlett-Packard security works on the breaches in Microsoft's Windows. The major problem in computer security is that people won't upgrade their software, allowing hackers to intrude on them.

According to a survey by Ericsson 2018, there are around ten devices connected to the internet on an average in an American home.[45] When we speak about companies, it is increasing exponentially. Most of the companies don't keep track of multiple systems they own, allowing hackers to invade into confidential data.

There is an interesting notion out there, which has proposed getting rid of passwords altogether—users hate them, hackers love them, and security staff dread them. According to Verizon, if the victimized firm needed more than one password to join its network, a quarter of the information breaches analyzed could have been prevented. The hackers looked at targeting organizations with enormous caches of passwords. This is because the hackers found out that individuals tend to use the same email address and passwords for different accounts from banking to social media. As organizations acquire more machines and build big networks, there is an increase in problems. Many computers will have administrative passwords accessible online as the default, and some organizations will not configure anti-hacker devices to monitor traffic in and out of certain systems.

The report by Ponemon Institute, '2019 Cost of a Data Breach Report' emphasizes on global financial risks, its underlying and mitigating causes of cyber breaches. The study gives information on the cost associated with the infringement at 507 organizations of 16 countries/regions, and in 17 industries. The 2019's global average cost of a data breach is USD 3.92 million, a rise of 1.5% from the 2018 study. The US is the most expensive country with USD 8.19 million loss on data breach, i.e. more than double the global average. Healthcare is the most expensive data breach sector with an estimate of USD 6.45 million in total data breach expenses in 2019. [46]

The life cycle of infringement is 279 days in 2019 with that of 266 days in 2018. However, the longer the life span of a breach, higher the total cost. The criminal activities require an average of 314 days to detect and destroy. Malicious and malicious threats constitute 51% of the leading root cause for data breaches in 2019. The other two root causes are system glitch infringements (caused by technology faults), and human error infringements (triggered by an individual's mistake and malfunction).

Verizon's Data Breach Investigations Report, 2019 projects: 43% of breaches involved small business victims, 16% were the breaches of public sector entities, 15% breaches involved healthcare organization, and 10%, the financial industry. 39% of breaches were performed by the organized criminal groups. 69% of breaches perpetrated by outsiders while 34% involved insiders. The tactics used for breach majorly involved hacking (52%) and social media (33%). The motive behind these breaches triggered monitory benefit with 71% and strategic advantage with 25%. [47]

According to CB Insights data, companies are trying to mitigate data breaches and vulnerabilities in data protection, as cyberattacks are becoming more common. In the year 2019, 23 cybersecurity companies exited with valuation over USD 1 billion. Those included McAfee—a security and antivirus corporation; CrowdStrike—an AI-powered endpoint security platform, the largest cybersecurity IPO; and CloudFlare—an internet security monitoring startup.

The future of enterprise cybersecurity is becoming defensive and firms are adopting radical cybersecurity approaches to go beyond network defenses. Financial firms are among the hardest hit by cybercrime than any other industry. In the year 2018, roughly

30% of all cyber incidents reported, occurred in the financial industry. Finance Industry is the first sector taking strategies of military and intelligence agencies to fight back against cyber criminals.

Multiple companies have reached unicorn valuations prior to exit, and those included Duo Security (USD 1.2 billion), FireEye (USD 1.1 billion), and Crowd Strike (USD 1 billion) etc. Post the global pandemic of 2020, Cybersecurity companies have become and even more critical part of today's businesses and commerce. The Work From Home requires broad based security across user end point management, access management including zero trust and identity management. Companies like Okta, Crowd Strike, Mobile Iron, Palo Alto Networks and Microsoft are leading the way. Cybersecurity startups are an attractive investment target. Private companies are stepping up, backed by millions of dollars of funding to solve the problems of mounting cybersecurity. In year 2019, the cybersecurity startups have raised a record 8.8 billion in disclosed equity funding. It is estimated that cyber coverage could increase from USD 7.5 billion to USD 20 billion by 2020. The surge in this potential growth has prompted the investors to contend for a position as well as create a host of new startups.

Data protection is essential regardless of the business entity. The probability of data breaches cannot be ensured by investment in protection alone. Companies should be able to detect the existence of data breaches, repair them, and avoid them at the same time.

11

Fintech
- Unbundling Banking and Finance

"Banking is necessary. Banks are not."
— **Bill Gates**

Like every other industry, traditional banks are under attack from a number of emerging specialists Fintech startups. Fintech players are focusing on improving specific parts of traditional universal banking models that involve broad product portfolio in commercial, retail, transaction banking, wealth management, investment management, insurance etc. The new age fintech startups are designing, developing and implementing scalable solutions, that are better, cheaper and faster than what the banks currently offer.

Today's fintech companies are innovating by focusing on the limitations of traditional banking processes. They are leveraging technologies like mobile applications, cloud technology, crypto technology, data analytics and artificial intelligence (AI) to enhance convenience, user experience and address the functionality gaps, which are difficult to bridge using traditional models. These technologies are enabling fintech startups to assess a specific part of banking and provide a superior alternative. According to KPMG, the investment in fintech market worldwide grew from USD 50 billion in 2017 to USD 111 billion in 2018.[48] According to Mordor Intelligence report, the global fintech market is expected to reach over USD 26.92 billion by 2024. This is clear evidence of the rapid innovation in financial services.[49]

The wide range of solutions offered by these new fintech startups are improving the financial processes and offerings, which are controlled by banks. A single new tech startup is definitely not a major problem for a giant traditional bank, but the threat lies in the number of such tech startups. The number of such fintech startups is now big and they are starting to take a large share of the market from the banks.

These highly specialized startups are taking advantage of the fact that no bank can be good at everything, everywhere and for everyone. In every single service that a bank provides, there are a number of fintech players, trying to take the business away. Battling for business with their traditional competitor banks has long been a part of the game for the banking sector. However, the scenario is increasingly intense, with all parts of the value chain being under threat, resulting in new evolving rules of the game.

A survey of 2,450 millennials from United States and Canada revealed that 67% of them are open to using non-financial services brand; while 46% say they don't plan to stay with their current financial services company.[50] In the past, customers had to rely on a big bank for all their financial service's needs—checking accounts, loans, insurance and wealth management. Here comes fintech startups like Opendoor, Acorns, Stripe, Credit Karma, Coinbase, Robinhood and others which gives control of distinct aspects of their finances to clients. These startups are also removing the friction of engaging transactions and lowering the transaction costs. Transferring money, paying bills and managing investments is now very simple and can be done with few taps on a smartphone. The barriers that new startups will face to enter the market has been significantly lowered after the first phase of unbundling. By the time new participants' rise in the market, a huge number of potential clients would have already swung to a non-bank technology company for their finance and banking needs. Fintech service will be considered as an elective, even by those clients who don't use them right now. A drastic shift in their own behavior is already evident and is expected by the clients over the next few years.

According to the' Global Fintech Adoption Index 2019' released by EY, the adoption of Fintech products has steadily increased from 33% in 2017 to 64% in 2019, which has become the majority of the surveyed market. More than 27,000 customers in 27 different markets participated in the study.[51] The aim of the survey was to gain a global understanding of trends in Fintech adoption among companies and demographic groups over time. *"The Fintech industry has grown up and grown out. No longer made up of only startups, Fintech today is a host of seasoned companies that offer a broad array of financial services and operate on a global stage."*— EY. The survey's key findings are that 96% of global customers are aware of Fintech's money transfer and payment service. 68% customers will prefer a financial services company and 46% were willing to share their bank information with other organizations.

People are ready and willing to consider fintech alternatives. 73% of consumers say that they would consider using technology providers for services that they usually use their banks for. It is common assumption that concerns about the security of non-bank providers have been a barrier in the uptake of fintech. However, there was little differentiation in the research between the views of early adopters and general consumers on this matter. 54% of early adopters—compared to 52% of general consumers consider technology providers as or more secure than banks and around a third (32% of early adopters and 33% of general consumers) consider them less secure. As people have started trusting tech companies like Google, Apple, Facebook and Amazon—eventually will start trusting startups too. If security is of equal concern to early adopters and mainstream alike, what is the differentiation? For 62% of early adopters, cost is the primary motivation for using technology providers rather than banks. The services that were first to be disrupted were those where the consumers are getting the rawest deals—the money saved becomes the reason for switching.[52]

Banking consumers use multiple services from a single provider but that is not what technology providers are offering currently. There is a tech solution that can meet multi services need of customers. The potential for the smart phone to act as the single gateway to multiple services is real. The interaction between different apps can allow the phone to take the place of bank.

Advantages of fintech is that–it gives controls in the hands of their clients and clients don't need to place their whole finance into the hands of a single company/bank. Eliminating the banks seems to be the aim of companies like Moven and Simple. This will be achieved by providing banking without any fees. Clients' account comes with debit card that can be used at thousands of ATMs and includes budgeting tools and money transfer capabilities. When BBVA (Madrid based bank) acquired Simple for USD 117 million, it had 100,000 users.[53] BankMobile is an example for a new participant in the fintech industry, having intentions of targeting millennials with their suite of mobile friendly features. 2CheckOut, Paypal Payments Pro, Amazon Payments and PayU are some of the services that allow consumers to pay for goods and services or transfer money to friends and family. According to Statista 2019 report, PayPal has over 286 million active customer accounts, and the service's annual mobile payment volume amounts to USD 227 billion.[54]

The wealth management sector is also being hit by the unbundling of the bank. Consumers take control of their investment portfolio through digital channels rather than giving full control to the investors. Betterment, Carta, Wealthfront and Robinhood allow consumers to invest their money where they want it, with free stock trades, portfolio management tools and automated investing based on consumer's goals. These robo-advisors have gained lot of popularity among millennials and drawn attention of large banks. Wealthfront, the automated advisory is gaining popularity in wealth management. It is one of the biggest players in automated investment advisory having over USD 11.4 billion in assets under management. Wealthfront has competition from the likes of Personal Capital, Betterment and Bambu.

Many times, a very high fee is charged by credit card companies from businesses to accept them. There have been instances where credit card companies like Visa and Mastercard had worked together to hike the swipe fees, which led to a lawsuit being filed against them in 2005. Square and Braintree are some services that allow small businesses to seamlessly accept payments via credit card, bank transfer, and even bitcoin with significantly lower fees and less red tape.[55]

Services like CAN Capital and Kabbage provide business loans and merchant cash advances for small businesses, while companies like Fundera match companies with lenders and offer the most competitive rates. Zenefits, Wave and ZenPayroll are helping small businesses, that can't afford a dedicated accounting to ensure on-time payments to employees and vendors.

OnDeck is a non-bank small businesses lender, based in New York. The startup grabbed investor's attention that raised USD 180 million in equity financing and USD 300 million in debt.[56] Businesses are opting OnDeck over traditional banks because OnDeck is faster and the loans can be processed within days. Sometimes banks take couple of week time to make a decision.

Currencycloud is an international payments engine, which raised a total of USD 80.2 million in funding over 10 rounds.[57] According to Mike Laven, CEO of Currencycloud—this software as a service company has become Amazon AWS of foreign exchange and has over 350 companies using its services to build their own cross-border payments offerings. Currencycloud offers transactions across 212 countries in 35 currencies around the world.[58]

Some fintech startups like Project Crowdfunding sites, GoFundME and Andreessen Horowitz are benefiting from existing technologies such as social networks and mobile messaging. Companies like Sentient Technologies and Ayasdi are using AI technologies to bring their products to fintech, for applications ranging from quantitative trading and sentiment analysis to threat detection and risk analytics. The appetite for using AI in financial services and telecoms has shown exponential growth in recent years.

IBM Watson is used by Barclays to provide services like money transfer and other rudimentary tasks. Bank of Tokyo–Mitsubishi, employs a 58 cm tall robot, named Nao, to perform reception duties for visitors, offering prerecorded responses in a number of languages, dealing with requests effectively and delighting customers as a result. One of the first robots developed to live with human beings, "Peper" can act by recognizing emotions. It greets and interacts with customers at SoftBank Mobile, a large mobile phone operator in Japan. Financial services have been revolutionized by the computational tools in the last two decades. Computers are now able to crunch more diverse and deep data sets than ever before by leveraging on technologies such as big data, algorithms and machine learning. Financial services companies are witnessing a new era of possibilities with advent of more powerful computers at low prices, social networks, mobile phones and wearable devices.

Kasisto is pairing user experience with AI to scale the impact of humans using technology. AI technologies are being utilized to deliver human like chat experience in customer support segment without the need of nearly any human assistance. A combination of UX (user experience) with smart agents that can analyze and collect data about customer behavior and compare to broader datasets can enable companies of all sizes to deliver personalized financial services.

UX integrated with AI has potential to deliver variety of digital fintech services like automated advisory, mobile wallet, personal financial, investment advice, online banking etc. The mix of technologies can enable companies to provide services to segments of customers where they were unable to provide high touch, human service. These AI powered fintech solutions can provide advice at the transactional level. Smart wallets like wallet.ai can assist consumers, analyze, price, and consider every single thing they spend money on, at a granular level that no human assistant could match. The idea of augmenting human interactions and intelligence with AI is not limited to consumer facing products. AI could also use power technologies that overlay humans to provide an oversight and tracking mechanism to employ actions. Computers can watch and learn over time to verify data entry and test for specific events, assess risk, and find fraud. Any fintech area that is regulated, provide companies with the opportunity to deploy AI-powered employees and oversight systems.

The unbundling of finance and banking sector should be a RED alert for banks. The newspaper industry is a great example. Newspapers survived the arrival of 24-hour cable news. However, with the invention of internet and advent of mobile apps, there has been a significant cut in number of newspaper readers. In effect, newspapers have been unbundled.

Similar trend can be observed in unbundling of retail banks. People are increasingly adopting specialist providers offering competitive price and service. The fintech startups now provide alternatives for almost all bank services from current accounts, payments, insurance, loans and savings. A BCSG study—

'Redefining Digital Banking for Small Business' reveals that the relationship between retail banks and their SME (small to medium enterprise customers) is weakening, and there is increased risk of switching financial service providers. Banks communicate either face to face or through digital channels. An astonishing 73% of SMEs have no contact whatsoever with a relationship manager. As a result, 67% of UK SMEs are now happy to look elsewhere for financial services and more than half are tempted to switch banks.[59] 2016 was a landmark year for Fintech, Financial companies across the globe raised about USD 36 billion in over 1500 funding deals from over 1700 unique investors according to data from Financial Technology Partners. The highest funding was poured into Payments/Loyalty/E-Commerce companies of over USD 13.5 billion followed by USD 9.3 billion into Banking/Lending.

According to KPMG's International global fintech survey 2018, the top three greatest sources of disruptions showed— emerging financial technologies at 57%, growing global regulatory complexity at 51% and new business models at 46%.[60] The other factors that contributed to sources of disruption included—increased cyber threats, increased customer adoption of mobile devices, competition from new entrants, labor and talent shortages, changing customer demographics, passive strategies and increased number of retail channels. According to KPMG, some of the top most innovative fintech startups in the world listed—Ant Financial (online payment services provider is the world's most valuable fintech company with 870 million users worldwide), JD Finance digital (provides online and offline services with user, data and connective as its key points), Grab (provides essential services to consumers like ride-hailing to super app), Du Xiaoman Financial (provides short term loans and investment services), SoFi (online finance company providing student loan refinancing, mortgages, and personal loan), Nubank (100% digital channels and reduced paper frictions).[61]

The key product developed at Tholons Digital is a fully digital insight based, ML/AI fintech platform for Know Your Client/Anti Money Laundering (KYC/AML). This fintech solution incorporated AML data analytics from Amberoon and intelligent automation from Blue Prism.

This was first beta created for a top 5 US bank with compliance audit review by one of the top big four. This solution addressed the entire KYC refresh of US consumers and Global wholesale banking. This solution enabled the bank to cover mid-risk clients that otherwise would have required them to quadruple their KYC team from 1200 to over 4000 bankers at a cost of USD 200 million/year. The Tholons Digital fintech solution enabled the bank to deploy at an average cost of only USD 100 million/year and with less than 400 bankers.

According to CB Insights data, there are over 20 Fintech startups that have crossed the 1 million customer account mark. In aggregate, these companies have added 200 million accounts over the last 10 years, and have raised USD 7.68 Billion in combined equity funding since their launch. Many of these companies are planning to expand into new products and markets in order to further grow and retain customers.

Fintech products are gaining the most traction with customers, and are starting to nibble business from the incumbents. 48 of the top 250 fintech companies are unicorns (2019) and have a combined valuation of approximately USD 142.17 billion. Fintech startups are changing the consumer experience and capitalizing on a growing consumer demand for more accessible financial services. To stay competitive, startups are iterating quickly and offering new products and tailored services ahead of incumbents.

Rather than competing with the new age technology startups, banks should consider establishing themselves as a vital part of digital fintech ecosystem. Banks must embrace new fintechs and other tech startups. This will help banks get deeper insights about customer needs, enabling them to offer relevant products and services that deliver amazing customer experience and value.

12

Healthtech
- Redesigning the future of staying healthy
Author: Dr. Garima Sathi Vashistha

"Firstly, the idea of receiving medical treatment exclusively at a doctor's office or the hospital will seem quite quaint."
— Harvard Business Review

Innovation is targeting every corner of the industry, and healthcare is one among them. Innovative startups and companies are offering wide range of technology enablers for clinical solutions, healthcare applications, medical devices, wearable technologies, data analytics and virtual reality. These enablers are key disruptors in healthcare delivered.

Adoption of IT in healthcare has followed the same trends as other industries. The beginning of the use of IT in the 1950's to automate repetitive tasks like healthcare payers, accounting and payroll can be considered as first collaboration of technology with healthcare. Second revolutionary wave for IT adoption in healthcare arrived in 70's, which supported B2B processes such as supply chain management within and outside individual industries. The effect of second revolutionary wave of IT adoption in healthcare industry was seen with the introduction of electronic health card in Germany. It was also a catalyst for the health information technology for economic and clinical health act in the United States and the national programme for IT in the National Health Service in the United Kingdom.

Healthcare systems are bound to resolve a host of challenges and leverage opportunities presented by medical advancement. The advancements in genetics, IT and nanotechnology have given rise to a more personalized healthcare approach. The healthcare system is evaluated using parameters, such as access to the best available treatments and to non-institutionalised care; compliance with treatments and even patient's choice. This highly dynamic healthcare environment has created a need for companies in healthcare to start thinking about what the future of healthcare might look like. Many are too constrained by their standing status quo of how the industry has operated in the past, or they have a rather narrow perspective. Many believe that due to the sensitive nature of healthcare, patients will not opt for digital health care services. According to McKinsey's global survey, more than 75% of people would choose digital healthcare services for its services that meets and provides expected level of quality. It also shows that patients of all age groups are ready to use digital services for healthcare. More than 70% of all older patients in the United Kingdom and Germany wanted to use digital healthcare services. In Singapore, the number is even greater.[62]

To most people, the 'robotic doctor' sounds like a thing of fiction, but it plays an important role in the future of healthcare. The innovation in technology is already shifting the perceptions and shaping the new reality. Data companies are trying to revolutionize the healthcare industry. New trends are starting to emerge in the rigid and stubborn structure of healthcare. Due to the availability of data on cloud, healthcare tasks are moving online. Digitalization of diagnoses has helped to prevent diseases by providing necessary assessment and medication. For instance, detecting the early symptoms of a virus or increase in body temperature will help people to take preventive measures.

The devices linking personal fitness to an individual's lifestyle and supporting predictive medicines have become a necessity. Entrepreneurs are utilizing smart devices and vast amount of clinical data to improve the patient experience. They have done advancements in big data and predictive analytics that led to creation of robots doing healthcare tasks on their own. For example, Eve is a robot, built to deliver vital samples and supplies at San Francisco

Medical Centre. Entrepreneurs must collaborate with healthcare organizations and other key healthcare players to address the gaps in technology. This need has already been recognized by incubators like Chicago's Matter, the incubator which connects innovators with research institutions and other health care stakeholders to create new age life-science startups. High tech devices have already become a part of healthcare systems and playing a major role in improving the quality of healthcare.

There is already a very diverse and exciting group of sensing technologies available, that require the patient to provide just a single drop of blood mixed with nanoscale test strips for process. This identifies the ailments ranging from simple colds, flu and more serious diseases like Ebola, with gold standard accuracy. A wearable patch monitors vital signs, such as breathing and heart rate, sharing data over Bluetooth with either the device or the user's smartphone.

Those who will adapt and innovate technology to connect doctors with diagnostic tools, will be in a position to provide scalable supply to meet demand. The most common deliverable in healthcare is to provide assistance during daily tasks and navigating the often-complex healthcare system. Countries with different healthcare systems cited one common assistance requisite that patients' most often look for help—finding and scheduling physician appointments with prescription refills and selection of the right specialist.

There are many inventions by startups in the field of healthcare. The applications help diagnose, monitor and manage patients in a better way. The wearable technology has made an impact on most of the industries enabling patients track data. Dr. Rafael Grossmann was the first surgeon to use google glass while performing a surgery. This wearable technology is meeting the demands of patients as well as doctors.

The opportunity to innovate the way clinical trials were conducted on animals, led to the invention of microchips modelling clinical trials. This technology enabled the scientists and doctors to conduct the trials, safely and with precision for better human patient treatment. The microchips developed can reconstruct the complex interfaces between organs and capillaries. This idea is similar to the idea of micro fabrication. It has resulted in reconstruction of organs by focussing on the use of complicated system of microchips. 3D printing is another innovation in healthcare sector. Today, the technology has large number of applications varying from art and design to a complex architecture model. The medical uses of 3D printing are extremely practical; the technology can perform tasks like printing human skin to blood vessels and heart tissues.

There are so many companies working to advance healthtech. As technology evolves, consumer driven healthcare becomes more appealing. Companies like iCouch, a mental health technology company provides secure communications, scheduling and payments for mental health professionals. It is taking the world of psychiatric analysis by storm. It can be a hassle to drive to a therapist's place back and forth. iCouch is a web app that pairs users with therapists who typically charge between USD 65 and USD 90 for 50 minutes of video chat time. The app, iCouch.me can be used via web on webcam enabled desktop or laptop. iCouch creates partnership with hospitals, insurance companies and hospitals to offer mental health services to the respective partner's employees, customers, and patients. It has more than 5000 paid users, and is adding about 15 therapists per week to its database of several hundreds of them from all over the world.

Instead of using paper, healthcare organisations are going for Electronic Medical Records (EMR). Practicefusion, a cloud-based service which can be set up in minutes, is the fastest growing EMR community in the US with over 150,000 medical users who have access to the health records of 33 million patients. Hospital records and processes are moving from written to digital. Most of the people use their mobile devices to do work. Even physicians are taking that route. Doctors use platforms like DrChrono iPad Patient Care Platform. It helps them to manage their practices with help of different electronic devices. Activities like electronically prescribing, billing any insurance company or maintaining EMR are all being carried out on the platform. The company has seen exponential growth, going from a small user base of hundreds to more than 15,000 registered healthcare providers and more than 400,000 patients.[63]

With all new technology coming into the market, speed of information exchange is becoming the key. With the development of SwiftPayMD, doctors can dictate diagnoses and billing cost.

Benefits of using SwiftPayMD are no delays, correct billing amount and time saving. It is available for USD 99 per month.

The healthcare industry is expensive, and people pay a lot of money to fund their well-being. Simplee, is an American software-as-a-service company that develops payment software for the health care industry. It helps people to manage their out-of-pocket healthcare costs which, on average, run about USD 3600 a year for a family of four. Information like money spent on healthcare, health saving account balance and status of insurance deductible is readily available on the platform. Since launching in 2011, Simplee has tracked hundreds of thousands of doctors' visits worth USD 200 million in claims.

Doximity, like LinkedIn, connects doctors on a social media platform. It gives doctors the privilege of consulting another doctor to seek help for a complex case. Doctors on Doximity provide information about their training, clinical trials they have conducted, etc., all of which will be accessible to all. Doximity members get paid about USD 250 to USD 500 an hour to hand over their opinions on such matters.

Robotics has come a long way in the healthcare industry. They are now a part of routine surgeries and work alongside humans to make life better. Robotic exoskeletons, in healthcare allow paralyzed people to walk. *"We are starting with soldiers and paralyzed people because their needs are great and opportunity for funding is better. But you can imagine exoskeletons for workers using heavy tools to hold for more than a few minutes. And a consumer version for people who want to run a marathon or climb Mt. Kilimanjaro."*—Eythor Bender.

As the digitization of most of the B2C industries has evolved the notion of healthcare consumers, the application of technology allows video conference medical consultancy. The new age smartphones connected with IoT healthcare devices are allowing doctors to monitor patient's health from a distant place. These devices are enabling doctors to trace activity logging, heart rhythm and many other in-depth physical examinations of patient. Enterprises with specialized knowledge and advanced skillset in areas like analytics and healthcare delivery will interact and compete through the common data and resource aggregation platforms.

Infrastructure providers will be providing the routine healthcare facilities.

Advancement in technology is not only helping healthcare providers in increasing patient engagement and enhancing service quality, it is also enabling governments to reach people with health care information, businesses to customize employee wellness programs and people to take control of their own health.

PwC's health research gives a fair idea that people are ready to abandon the traditional healthcare techniques. Activities like examining strep throat or chemotherapy should be done at home. According to David Moen, a physician and medical director of care model innovation at Minneapolis-based Fairview Health Services, the future of care includes different delivery models like phone, internet and group visits. As part of Fairview Health Services pilot program with BlueCross BlueShield of Minnesota, it started to see some patients via internet using a webcam and a telephone. This online care model does not only extend healthcare access to people in rural or underserved areas, but it can offer the physicians a unique way to control schedule. Technology has provided caregivers ways to continuously engage with patients for providing excellent healthcare from anywhere. It is also helping patients and providers by provides a much necessary feedback loop.

Telehealth, remote biometric monitoring, and technology assisted health coaching are powerful tools to improve chronic care outcomes. They support patients in learning to manage diet, exercise habits and medication routines. Type 2 diabetes patients were provided with a mobile app to motivate them and give behavioural coaching. This led to drop in patients' A1C level (measures the glucose level/blood sugar level in the blood). Companies have rapidly started adopting technology for their wellness programs. For instance, Virtual Wellness portal extracts numbers from on-site health screenings and provide recipient with action plan.

There are options available to examine moles, rashes and other skin diseases with digital photographs. These platforms also allow consumers to connect to a specialist of choice for online consultation. CellScope's Oto, a smartphone accessory that captures digital images of the ear canal, went on the market at the end of 2014 at USD 79 for home users and USD 299 for medical professionals.

Scan-adu scout, a replica of "tricoder" of Star Trek, hoped to be part of every family's medical kit.

The Global Healthcare Market valued at USD 8,452 billion in 2018 is expected to grow at USD 11,908 billion by 2022[64]. This accelerated growth will be derived from several factors like government support for healthcare IT solutions, regulatory compliance management, high return on healthcare IT solutions; inclusion of big data, virtual reality and internet of things etc.

The telemedicine market valued at USD 21 billion in 2018, is estimated to be valued at USD 60 billion by 2024, a CAGR of 18.50%.[65] It has an immense potential to play a major role in meeting healthcare needs of growing global population. The exponential rise in population has given rise to chronic diseases. Factors such as scarcity of healthcare professionals' worldwide, need for affordable treatment options and improvements in infrastructure gave rise to telemedicine market. Data analytics impacted segments like value-based patient centric care, where care providers, doctors, hospitals, and health insurance work together to deliver personalized care. Patient satisfaction is measured basis the care that is efficient, price conscious and transparent.

The advancements in digital healthcare have put traditional healthcare techniques at risk. But they still have a way to save themselves by collaborating with the new age players, gain knowledge on implementing these technologies and applying effectively. Global Digital Health 100 is one of the Healthtech industries foremost technology awards programmes for innovation and entrepreneurship. It identifies and supports healthcare technology companies for its demonstrated delivery potential. Some of the innovators that applied technologies like artificial intelligence, virtual/augmented reality, telehealth, digital therapeutics and behavioural change programmes are: Acuity Link –It is a comprehensive communications and logistics management platform that links healthcare systems with non-emergency medical transportation (NEMT) providers and ambulance crew members for all levels of care and modes of transportation. With Acuity Link's technology, organizations benefit from instant access to the closest available and most suitable transportation resources, reducing bottlenecks that impact patient flow, leading to shorter discharge times, and in turn, enhanced patient care and experience. Ada Health–It offers an AI-powered health platform that is helping millions of people around the world understand their health and navigate to the appropriate care. Ada's sophisticated artificial intelligence technology also supports clinical decision making and enables payers and providers to deliver quality, and more effective healthcare. Ada launched globally in 2016 and has been the number 1 medical app in over 130 countries. Corti—a machine-learning company that provides accurate diagnostic advice to emergency services, allowing patients to get the right treatments faster. Corti can identify patterns of anomalies or conditions of interest with a high level of speed and accuracy. In the case of out-of-hospital cardiac arrest (OHCA), Corti can reduce the number of undetected OHCAs by more than 50%. Propeller Health–is a leading digital therapeutics company dedicated to the development and commercialization of measurably better medicines. Propeller creates products to more effectively treat disease and improve clinical outcomes for patients across a range of therapeutic areas through connectivity, analytics, and companion digital experiences. The Propeller platform is used by patients, physicians and healthcare organizations in the US, Europe and Asia.[66]

According to CB Insights, over 150 digital health startups have made its hallmark on the transformation of the healthcare industry, with new models and tech solutions. These startups have developed AI tools for disease diagnosis, health insurance, and drug discovery. The startup encompasses patients, healthcare providers, and insurance companies as primary stakeholders.

Medical devices are evolving beyond sight. Digital therapeutics 'Abilify Mycite' is an ingestible embedded sensor drug that is helping in expanding the scope of medical devices. Grail, a genomics startup received funding of USD 1.61 billion; Oscar Health, an insurance tech startup, received funding of USD 1.26 billion; We Doctor, a healthcare platform startup received funding of USD 1.03 billion in 2019. AI is in the forefront for consumer health companies in imaging, and diagnostics.

Understanding consumer's needs and wants, can be the foremost step to start with. New age tech-based healthcare providers are offering wide variety of services to choose from. Healthcare companies should focus on providing exemplary customer experience that can earn trust. If the demands are not met on time, patients will leave the healthcare providers. Making pricing and payments hassle-free and transparent as possible, aligning working hours to patient's convenience and making services available through digital devices, are few vital adoptions for incumbents. Despite all the advantages, there is still a need for old and new enterprises to be more creative. They must win patients' trust. Customers will win with more high quality, affordable and simpler options.

13

Retailtech
- Taking shopping experience to the next level

"The e-commerce industry is a force that no investor can afford to ignore."
— **Cushla Sherlock**

"To succeed in the digital business era, retailers have to stretch their boundaries. This requires rethinking their business model, deciding how to transform into digital businesses and making the right technology investments to become digital leaders."—Accenture. We are now in the time—the world of smartphone generation. Apple Pay, Google Pay, Storecards etc., are driving our lives every day. We depend on hundreds of cards and loyalty programs which can easily fit in physical wallet.

    The existence of department stores was threatened by the e-commerce wave, which hit the industry, that triggered re-invention of department stores for revival. Department stores have found new significance via different strategies: getting the rights to beneficial trendy brands, granting international expansion for international brands through in-store boutiques, increasing e-commerce impression to improve, and enhancing the store experience. Emergence of e-commerce and the rapid increase in the number of people on internet and smartphones has transformed the shopping experience. The demands of the tech savvy consumers from the retail industry are shifting from just high-quality products and service support to a personalized and seamless shopping experience.

Nowadays the success of a retail business very much depends upon its strategy to navigate the ever-evolving digital world.

"In today's increasingly connected world, brands and retailers are struggling to find ways to appeal omnichannel shoppers. Technology advances have created an environment in which the line between brick and mortar and e-commerce is blurred and fading fast."—Mike Paley, Former Executive VP of Shopper Marketing at The Marketing Arm.

Analysis of customer relationship management (CRM) to know more about the customer behavior, drive decisions about offers, and to allow personalization of business channels are the base of retail digital marketing. Now the marketing teams have expanded their skills in data analysis and digital marketing that were earlier considered to be IT team functions. Retailers are now developing actionable intelligence from industry generated data.

The retail businesses can benefit from the readily available infrastructure for capturing and analyzing their business relevant data. The expenses of transforming the retail business can be minimized using secure cloud services; these services come with a lower cost and offer greater flexibility. The retailers who embrace this new model will invest in data analytics, and IT skilled professionals, who can innovate the business into a data driven model.

Payments are one of the most rapidly transforming areas in retail industry today. With the invention of mobile wallets and e-payments, the retailers look for the ability to accept all payment options chosen by the shoppers. But, legacy systems, investment, agreement with payment processors, and the amount of time to establish these payment connections to their systems are the challenges faced by merchants, retailers and payment providers in accomplishing it. There is growth in number of modes of interface between shoppers and retailers in the recent years. This growth will continue to increase. Hobbyists, entrepreneurs, and even fans of specific product know that data is available online and people crave for that data. An interesting example of non-retailer developed channel is "The Liquid Control Board of Ontario (LCBO)." They are one of the largest alcohol retailers in the world. They have a wide variety of products and huge number of consumers. The app provides the price and availability of products to the customers. LCBO also has attractive offline stores.

Year 2010 to 2016 had retailers focus on either defining or evolving their digital strategy on the primary focus area—IT solutions, web content promotion, email engagement, social media sites and developing/enhancing mobile applications for digital branding. This focus yielded many benefits and gave retailers the capability to pursue digital strategy and drive consumer choice. In order to target audience, retailers follow different approaches, depending on customer endpoint and technology supply. They are trying to merge these experiences into a single digital channel which will dramatically enhance the consumer experience. This helps retailers to set up a cross channel platform. This is a retail revolution in the making, enabled by a digital platform strategy that allows retailers to co-create value across channels, partners, and potentially other industries or even competitors. Secure networks, cloud, and increasing mobile and internet penetration, have brought retail industry to realize omnichannel engagement.

The digital advancement has given rise to a cohesive consumer driven supply chain model which is characterized by flexibility and choice of fulfillment. In reality, consumers end up becoming a brand unto themselves. This requires retailers to adopt a new model of engagement with consumers to create a valuable supply chain. This flexible fulfillment supply chain will cut down retailer's costs of delivery, overstock and waste. IoT devices in retail tech are engaging and attracting consumers in more intimate ways resulting in generation of greater volumes of data than ever. With the help of IoT, retailers are automating many functions, which permit them to design marketing and target customers with greater experiences.

Social network websites are more than just for spreading the word. Big social network companies like Twitter, Facebook and Pinterest have all experimented with direct "Buy Item" tabs on their website. Twitter teamed up with Shopify, which had about 100,000 merchants, and other e-commerce software companies. Pinterest's "Buyable" button allows buying without leaving the Pinterest app.

Digital transformation is seen in in-store retail too. Retailers are leveraging in-store technologies. For instance, retailers such as

Ugg Australia (an American footwear company and a division of Deckers Brands), Uniqlo (Japanese casual wear designer, manufacturer and retailer) and Neiman Marcus are using "magic" or "memory" mirror technologies, using RFID (radio frequency identification) tags, which allow customers to try on virtual outfits in different colors and styles. Rebecca Minkoff has added text messaging and touch screen features in her stores that allow consumers to order drinks, browse the store catalog and easily interact with store associates. Bloomingdale is using iPads in fitting rooms to let customers ask for help, read reviews and see the size availability.[67]

Startups are innovating each and every aspect of retail store from deploying robots to streamline product, shelving to leveraging sensors for capture of foot traffic and customer behavior data. Startups offering point of sale financing options, displaying product information on digital tablets and more. Offline retail stores are increasingly collaborating with tech startups to stand a chance in competing with their online counterparts. InteractionOne is one such startup that enables user smartphone to buzz and directs to the shop that is close by to the user's physical location, basis the search that was performed earlier— "Nike shoes you looked up the other day is available in your size and color with 15% discount, at a store 300 meters away from you." Using machine learning algorithms, Moberry App traces the likes and preferences of users through their browsing patterns and allows them to make a wish list. The app then sends notification to the user's smartphone when the user is near a store. Once the user enters the store, beacons mounted in the store detect the user's presence and credits the user with walk-in points. *"Retailers have started to see traction and are more willing to integrate technology in their store."*—Krishna Prasad, Co-founder of InteractionOne piloted their first project based on IoT in Bangalore on a high-end shopping street with 80 showrooms.

A Nike Facebook app can be considered as a step forward, in enhancing customer experience; the app lets runners pay for kilometers completed. We are seeing a lot of apps nowadays that reward their users in many ways like motivational phrases for the users to attain fitness. Nike Mexico and Facebook took this to the next level by launching 'Subasta de Kilometros', which allowed runners to score points for every kilometer they run and then use these points to bid on Nike branded running gear in an auction.

In-store virtual reality technology is rapidly being adopted by Supermarkets in China to provide a more interactive shopping experience. Yihaodian developed augmented reality stores that can only be accessed in certain public locations like parking, parks, malls and other tourist spots. When customers point their phone in right direction, a virtual store is displayed, this provides a simulation of real-life retail store to give a more interactive online shopping experience. Lowe's, the home improvement store, claims that the customers who interacted with the 3D objects in its app had 104% higher conversion rate from those who did not. Augmented Reality (AR) customizes in-store experiences with mannequins that match the user body type and display enough virtual inventories to rival any online site. Merchants create promotional clips of AR experiences that emerge when the customer looks at the product on the shelves and enable celebrities to pose in the aisle to pitch the product. In Denmark, SuperBrugsen is using their website to ensure that the produced stock will appeal to eco-minded customers. Customers are welcomed to give suggestions on what local items they would like the store to stock. Based on the number of suggestions received about a particular item, an executive from store checks the item to ensure that the product meets their quality standards; a clever way to use customer crowdsourcing is to ensure that the store only stocks items that will sell. In the Netherlands, the crowds were also put to good use through the "Avoid the shopping crowds" feature which analyzes updates from social media to tell the user how crowded a shop is before they visit.

The retail industry is impressively adjusting to the highly connected world we are living in. Brazil based C&A company has shown a unique collaboration of the real world and the online one. The company displays the Facebook likes of their articles on a small screen embedded on the hanger of articles. The expectation was that the endorsement by online community would motivate shoppers to buy an item of clothing.

For giving a better shopping experience, retailers must take care of shopper's varying needs. Retailers don't have to depend on individually identifiable information to personalize shopping experience. Information such as items on shopping list, keywords selected, location, search items, and local store popularity are readily available for personalization. Retailers that can significantly leverage mobile technology will get a head start among competitors.

The increasing adoption of IoT in retail sector is driving the growth of retail market; the global connected retail market size is expected to reach USD 53.75 billion by 2022–Grand View Research Inc.[68] IoT is equipping retailers, majorly in supply chain, new channel, revenue streams and customer experience. The hardware retail segment is expected to grow at a CAGR of over 19% by 2022. The software retail industry growth will be driven by the increasing number of applications and software to support the IoT and other technology adoptions. The low power consumption bluetooth devices are expected to witness growth of CAGR 25% by 2022. The evolution of advanced barcodes and imaging technologies for scanning products, faster alternative checkout methods will continue to grow across the retail sector.

Social media platforms generating tons of customer data will be playing a vital role in marketing and sales, photo sharing and hashtags have become the word of mouth of the digital age marketing. *"Instead of having to leave the page or site to search and buy the products they see in photos, consumers will be able to easily click on the content, and purchase the products within, during that moment of discovery."*—Jim Rudden, CMO at TalentGuard.

Technology is playing a major role in training and empowering store employees. Store and product information are made easily accessible to associates through mobile applications, which help shoppers in-store to increase confidence and efficiency of retailer. *"Retailers can use mobile or retailer-specific apps to elicit customer feedback on interactions with sales associates, using the data to rate employee performance."*—Harvard Business Review. Keeping this data transparent to employees, retailers can maximize their performance and identify any stores which can improve customer service, helping to quickly course-correct.

Major retailers such as J.C. Penney, Macy's, Sears, Ralph Lauren and Kohl's are suffering disappointing earnings and shutting dozens of stores. In January 2016, Walmart announced closing down of 269 stores globally, with 154 of them in the US. This only accounts for a fraction of the company's 11,600 stores worldwide. In January 2018, without notice Walmart closed 10% of all Sam's Clubs. J.C Penney closed 139 stores in 2017. Sears had 766 stores as of November 2018 and planned to close 241 in 2019. Macy's department store has closed more than 120 stores since 2015; this represents about 15% of all Macy's department stores in the United States. Sports Authority decided to shut down all 450 of its locations after going bankrupt. Toys R Us, a darling of pregnant mothers, children and many others, is unfortunately out of business! Even the bleakest expectations aren't met by many. Gap's CEO stated that his company would be delusional not to start selling some of its apparel on Amazon and other sites. Amazon's domination in online retail sales is no big secret. According to Business Insider report, about 53% of all online purchases in the US went through Amazon.[69] The rest 47% were from all other retailers combined.

On the off-chance that organizations envision development in the age of digital disruption, it needs to get prescient consumer capabilities with innovation. Retailers are able to quantify the right economic value of their stores, crosswise over channels, that results in incredible insights. This is made possible through cutting-edge geospatial analytics.

According to CB Insights data, the retail ecosystem has evolved to stay competitive in the face of e-commerce and a rapidly shifting consumer landscape. The companies are implementing steps to evolve by targeting new demographics, providing discounts, cultivating new niches, repurposing the physical store and automating processes.

The rise of e-commerce is distorting both physical and online retail. Traditional retailers are the victim of the retail apocalypse. Startups have developed in-store technologies from shelf-stocking robots to payment systems. For example, Macy's expanded its partnership with virtual reality startup Marxent Labs, that allows customers to visualize furniture in their own houses.

RetailNext, a startup (USD 189 Million) offers traffic sensors, customer route mapping, and mobile marketing. Teemo (USD 17.9 Million), a performance marketing platform for physical retailers, combines user offline behavior understanding and algorithms to drive in-store visits, that are measurable. StartupTrax

(USD 272.6 Million) offers visual monitoring of products on store shelves, with the help of in-store cameras, robots, and mobile phones to help retailers optimize the organization of their stores.

Technology has led its way from using traditional devices to highly sensitive gadgets. Robotic Process Automation (RPA) is one such technology that allows the retailers to maximize their work, and to conduct the sophisticated analysis too, just like human workers would do. 55% of store processes can be automated by use of Digital Workers paving the path to Intelligent Automation.

14

Consumertech
- Emergence of Vertical Market Places

"The customer's perception is your reality."
— **Kate Zabriskie**

We are all evidencing the fact–we don't work the way we did before. Everyone is looking for change and so should be the wavelength of the product/service providers. There is always this growing expectation from consumers; and technology is the answer to it.

Companies should do extensive research on the returns. There are companies in the market which deal in the production of specialized products, targeted at a specific trait of consumers. For example, consider a company that manufactures medical equipment which is of no use for the general public, then advertising the product to the public, is a mere waste of money and resources. Alternately, the focus should be on promoting the product in medical industry magazines and websites related to the medical field and conferences. This will help to reach the target customers.

When launching a startup/creating a marketing plan for an existing business, one of the first promotional decisions to make is between vertical and horizontal marketing systems. Vertical marketing appeals to people in a specific industry/niche, while horizontal marketing targets numerous sectors at once through clever strategies and partnerships. A horizontal marketing system focuses on broader audiences rather than narrowing down into specific niches. This means that companies with products/services that have a universal appeal are more likely to 'go horizontal'.

Businesses with a distinct focus on vertical marketing systems seek to reach a very specific demographic.

Companies should look forward to opportunities for expanding the business horizontally as well. For example, the Integrated Workplace Management System (IWMS) is an office tool that brings similar programs in one interface that is constructive and attractive to different industries. This product must be marketed to target specific industrial domains like real estate, finance, and energy. The wearable is another sector where vertical markets have come in. Apple helped people to feel comfortable with the idea of wearing tiny data-collecting computer screens on their bodies. With the launch of Apple Watch, more players hopped on this trend.

The recent advancement in artificial intelligence has increased its demand for being incorporated into different technology, services, and software. Googles' search engine, iOS' Siri, and Windows' Cortana are some examples of interfacing AI in various products. *"Consumer technology markets are being redefined by a new set of consumer expectations and values shaped by global economics, technology, and social change."*—Gartner.

What consumers value enough to pay for, how consumers' values are changing, and how technology and service providers should respond to this, to increase their sales and margins? The consumer market is shifting rapidly to online. The retail market might dominate for some more time, but customers want seamless experience while buying anything. Retailers should start taking initiatives to make the shopping experience impeccable and organized. Technology and Service Providers (T&SPs) should start researching in understanding the customers, whereas business analysts should track the behavior of consumers across different channels, reallocate the resources, and rebalance the priorities as required.

The consumer market has become more sorted. Providing complex user interfaces is no longer attractive to consumers. Customers tend to buy products with additional features, but the preference is towards simple and user-friendly products. Technology and service providers must focus on innovation while keeping statistics in mind. Demographics affect consumption pattern, which in turn affects consumer culture, values, attitudes, and expectation.

For middle-class people, price is not the only factor to choose brands, there is much more than that.

Apple expanded beyond its traditional consumer electronics roots with its entertainment business—Apple Music and Apple TV. In March 2016, Apple announced CareKit, an open-source platform that made it easier for developers to aggregate and share patients' medical information with their caregivers—all with consent. CareKit enables apps to help patients manage diabetes (One Drop), monitor depression (Iodine), track reproductive health (Glow), and record asthma symptoms (Cleveland Clinic). Apple's approach to health is to operate behind the scenes by helping researchers, patients, and developers to make use of the health data gathered by smart phones/gadgets.

Amazon is a giant player in consumer electronics. It just does not offer devices like fire tablets and TV boxes, but also provides access to its key featured service like Alexa, an AI assistant. This made the news at the CES 2017 gadget show by being built into everything from automobiles, televisions, smart speakers, wearables, and earphones.

Sony's artificial intelligence robot–Aibo (RoboCup) can find a ball, dance, and even recognize different family members through its very realistic OLED eyes. Its cameras are located in its nose and have another camera on its back that helps navigate its charging station. Sony included some brand-new AI technologies in Aibo, which is developing its personality over time. For instance, the more you pet this robot, the more it will like you.

Consumer Electronics Show (CES) is one of the largest consumer tech show that attracted over 180,000 consumers in 2018. The 2019 event showcased truckload of new gadgets by the world's top tech manufacturers. Some of the prominent players included Acer, Apple, Arlo, Asus, Audio Technica, Garmin, Google, HP, HTC Vive, Huawei, Lenovo, LG, Nissan, Nvidia, Panasonic, Samsung, Sony, TCL, and others. The 2019 theme focused on innovation of 8K HDR pictures from the existing 4K HDR picture and to debut 5G in the UK and US. Some of the key consumer trends identified in the event–smart homes, smart speakers, more intelligent wearable technology, self-driving cars, streaming services galore, etc.

To improve people's lives, P&G is up integrating cutting-edge technologies on everyday products and services. SK-II's Future X Smart Store is transforming beauty retail shopping making use of facial recognition, and gesture-driven 'Phygital' experiences, augmented by its proprietary skin science and diagnostics. Olay's Skin Advisor platform uses artificial intelligence to provide personalized skincare analysis and recommendations by analyzing selfies and a short questionnaire. The Oral-B Genius X toothbrush uses AI to recognize how users are brushing and provides personalized feedback. The new Heated Razor by Gillette Labs features a warming bar that heats up in less than a second and elevates the shave experience. Opté Precision Skincare System combines optics, proprietary algorithms, printing technology, and skincare in one device that scans and detects hyperpigmentation applying a healing serum to the skin. AIRIA is a smart home fragrance system that uses patented capillary action and heating technology to establish scent-enhancing ambiance with the touch of a button. *As consumers are changing, so must the FMCG giant. What remains the same is our focus on deeply understanding how consumers live, work and play, so we know precisely what they want. When we combine breakthrough science and technologies with this deep consumer understanding, we're able to deliver transformative innovations that improve life every day.*—Kathy Fish P&G chief research, development, and innovation officer.

The Savvy Home Smart Mirror by Electric Mirror allows mirror users to check weather and traffic, or even use it as an alternate screen to watch TV shows and order pizza. Delta's Touch2O, a digital technology that enables taps to follow verbal user instructions to fetch water in a tub, on and off water flow, run a bath while monitoring water usage. KitchenAid's smart oven is controllable through an app or by voice recognition. It identifies different ingredients and suggests recipes to help reduce food waste. Pepper, the US startup debuted a food scale that monitors calories, sugar intake, fats, and other nutrients, to help manage a healthy diet. Samsung's Bot Care is a health monitoring robot that detects breathing during sleep, follows heart rate, alert reminders to intake medicine, sense accidents, etc.

Customer Relationship Management (CRM) systems play a vital role in the growth of companies. CRM systems can predict consumer behavior, automate outreach, and promote interaction more efficiently with the help of leveraged data. According to CB Insights data, Industry Analyst Consensus predicts, the demand for CRM systems is expected to reach USD 44 billion worldwide by 2020. The management area for customer relationships generally encompasses companies that develop marketing supporting software and customer engagement services.

The way consumers shop was transformed by an explosion of new direct-to-consumer (D2C) businesses. In the process, these brands are radically changing consumer preferences and expectations, spanning everything from detergent to sneakers. Casper is impacting the furniture industry; Dollar Shave Club and Harry's are making their presence felt in the shaving industry, and The Honest Company is becoming a player to reckon with in laundry and baby products.

AI-powered Chabot's offer new ways to strengthen customer relationships with organizations. It is essential to understand and adapt to customer needs for long-term success. According to the consensus of CB Insights market research, the customer engagement space is expected to grow to USD 22 Billion by 2023. The focus growth area includes building conversational assistants, customer training tools, customer success analytics, sales enablement, and product feedback platforms.

The magic mix occurs when companies find a low competition sub-niche in the growing market. This assists organizations to get into the vertical market with little or no modification to their existing services, making use of less capital outlay, minimum risk, and maximizing the opportunity for success.

15

Edutech
- Nirvana of Learning

"Education is not the learning of facts, but the training of the mind to think."
— **Albert Einstein**

In a nutshell, modernization is the cause of making lives better, either a little improvement for one individual, or a boost for the society, "what's next?" is what makes us aim for a better future. On large scale basis, an innovation creates a change that alters all forms of our lives—the changes in the way we travel, communicate, live, learn and think. The introduction and acceptance of innovation like railroad, car, airplane, radio, television, computer, internet, robot and automation have changed our perspective of the world—so is technology driven edutech. Student must become an autonomous and tenacious starter; crave to possess technical or specialized skills for a fascinating future.

Policymakers and administrators are testing with new schools. Students are clinging to online learning, adaptive games and coequal learning platforms. Teachers are developing better teaching tools, content, pattern and application of modern research and technology to adapt and refine advanced methods. Business visionaries are making both for profit and for social cause. Startups are addressing on the areas like childhood improvement, tutoring, data analytics, student loans, alumni networking and almost everything that is most compelling direct towards refined learning solutions.

One of the biggest barriers is the traditional report card which is not as good as it sounds. It comes out with grades that don't deliver parents adequate data for assistance to their kids on the journey of attaining quality education. Ingenious learning requires inventive execution and application of thought process. Many schools resort to a scripted curriculum that is taking away teachers' expertise and autonomy. Most people carry around traditional views about what constitutes learning. Students are taught in a system that centered intensely on learning actualities. People still incline to relate the state of being well taught by the knowledge of abundant facts. Finally, the educators judge the advancement of students in terms of what they know. The need for top down change programs from the district and state level can certainly support educators towards lasting innovation from collaboration amongst learners, educators and communities.

Clayton Christensen, the leading expert of disruptive innovation, composed two books particularly centered on education. According to Christensen's disruptive innovation theory, markets are disrupted when new entrants figure out an innovative way to provide a simpler product to a wider set of buyers at more affordable price. Since the less complex product is really what the broader markets choose, the new competitors change the market behavior considerably.

There has been a greater acceptance of online learning and in parallel, it is noted that there is a rise in burden on education budget across the world. However, there is continuous effort in education sector to hike its investment in innovation. The global e-learning market in 2018 stood at USD 190 billion and, is expected to hit over USD 300 billion revenue by 2025 according to Global Market Insights Report. Digitization in APEC, EMEA regions and economies like the US adopting the education technology has led to this phenomenal growth. According to market research, Asia is one of the fastest growing learning markets in the world, and by 2020, it will hold 23% of the global Edtech market—Technavio.

According to HolonIQ, educations spend for the year 2018 hit USD 6 trillion worldwide and are expected to grow up to USD 8 trillion by 2025. Advanced education technology expenditure is in the process of hitting the stride with AR/VR, AI, Robotics and

Blockchain with about USD 4 billion in 2018 and expected to reach over USD 21 billion by 2025. The global education venture capital invested USD 2 billion in 2014 with that of USD 8 billion in 2018.

US has evolved as the pacemaker trend setter and has seen highest growth in Edtech market. Asia is encountering world's quickest development in investment into the sector, and Europe has seen increments in M&A. The growth of the global population will be the biggest challenge for education industry in the coming years—by 2035 there will be 2.7 billion students worldwide and the legacy education model will fail to deliver quality education to these growing numbers. Mobile penetration plays a major role in designing and delivering education in the near coming future. Coursera is one among many startups leveraging mobile technology to deliver specialized courses, offering Massive Open Online Course (MOOCS). It has recorded over 36 million users and over 1400 enterprise customers. It has partnered with 150 global universities and has launched over 3000 courses, of which, about 250 are specialization courses and about twelve, online degree courses. Coursera's massive scale model proved to be highly competitive and the revenue of the company reported USD 140 million in 2018 and entered Unicorn territory in 2019. *"In a rapidly evolving economy where skillsets continue to progress at an accelerated pace, digital models offer a way to capture these changes and offer new routes for re-skilling as well as addressing widespread labor shortages. 50% of current jobs won't exist in 2025, and consequently, there will be a growing need to re-train the workforce in order to address current skill gaps and increase the use of continuous learning"*— Benjamin Vedrenne-Cloquette, co-founder, EdTechXGlobal.

Online proficiency-based learning opportunities, such as online degree programs offer assistance to students through targeted learning outcomes, customized support, and portable skill sets that the corporation care about. This anticipates a tendency to showcase the act of company to establish a value network that helps students' interface directly with possible job offers.

The global higher education, testing and assessment market 2018-2022 witnesses an increase in the adoption of digital badges. Many higher education institutions are adopting digital badges in their assessment methodologies. Digital badges are a graphical representation of the performance of people. They are a validated indicator of the accomplishment, skill, and quality of students in both academic and non-academic learning environments. Using digital badges, schools and universities can measure and quantify the skill levels of students and their achievements.[70]

In order to keep pace with ever changing learning technology, online corporate learning is trend setting. The Clayton Christensen Institute foresees increasing propulsion for online corporate learning initiatives. Increasing number of companies realize that there's great value in furthering their employees' knowledge in ways that are flexible, cost effective, and tailored to each individual's needs. Enterprises centered on corporate e-learning have discovered a business model that is working well. The power of custom-made training from online corporate learning on any device are being experienced by industry employee at all levels, 24/7. The mixture of online learning and aptitude-based training procedure has made and is making a progressive approach towards education. It not only consolidates right learning model, but also the right innovation methodologies. *"Because that's what careers will require, education will be not just taking information and sharing it back, but also figuring out what to do with that information in the real world."*—Josefino Rivera, Educator in Buenos Aires.

"The classroom will be one big makerspace. Technology like Evernote, Google, and Siri will be standard and will change what teachers' value and test for. Basically, if you can ask Siri to answer a question, then *you will not be evaluated on that. Students will be* evaluated *on critical thinking and problem-solving skills. Literature and math will be taught, but they will be taught differently. Math will be taught as a way of learning how to solve problems and puzzles. In literature, students will be asked what a story means to them. Instead of taking tests, students will show learning through creative projects. The role of teachers will be to guide students in the areas where they need guidance as innovators. How do you get kids to be innovative? You let them. You get out of their way."*—Nicholas Provenzano, Educator in Michigan.

The evolution of adaptable learning tool from the traditional computer class has changed the method of how the projects are assigned, concepts are explained, and improvements are made. In a classroom, to explain concepts that are either big/small, or mechanism that are taking place quickly/slowly, digital models and simulations help teachers to illustrate in an easy way.

The free open source software that teachers can use to model concepts is developed by Concord Consortium, a non-profit that develops technologies for math, science, and engineering education. The Molecular Workbench provides science teachers with simulations on topics like fluid mechanics, gas law and chemical bonding.

"You used to count blocks or beads. Manipulating those are a little bit more difficult. Now there are virtual manipulative sites where students can *play with the idea of numbers and what numbers mean, and if I* change *values and I move things around, what happens."*—Lynne Schrum, author on schools and technology.

Technology is putting students into real world situations through innovative games, such as Epistemic Games. These games ask students to deal with real world problems making them act like journalist, city planner or engineer. The Epistemic Games Group has given a few illustrations of how students are involved in the adult world through commercial game like simulations can help students learn important concepts. *"Creative professionals learn innovative thinking through training that is very different from traditional academic classrooms because innovative thinking means more than just knowing the right answers on a test. It also means having real world skills, high standards and* professional *values, and a particular way of thinking about problems justifying solutions. Epistemic games are about learning these fundamental ways of thinking for the digital age."*—Epistemic Group.

Inventive educationists like Warhol and Escher—capture ideas and concepts from exterior space and discover ways to use those ideas to develop their work. They are utilizing their expertise in education and are developing solutions that challenge the existing status.

A lot has happened since 2005 or so in the effort to use technology as a medium to educate and train people. Devices, infrastructure, connectivity, services, products, content, marketing, parents, students and teachers are in line to adapt and revolutionize e-learning. *"Just like a mobile phone increased the connectivity exponentially from 3% that we achieved with landlines to almost 100% today, a smartphone is here to take the 'computer' penetration from 12-13% to around 60-70% in another couple years."*—Cisco.

One of the best education startups in the world is backed by leading venture capital firms such as Kleiner Perkins Caufield & Byers, New Enterprise Associates, GSV Capital, International Finance Corporation, Laureate Education Inc., and Learn Capital. Helphub—the instant tutoring platform enables teachers and students to work together using mobile and web devices. Its ability to provide an in-depth analysis of all the interactions which takes place on its platform makes it more popular among the student and teacher community.[71]

Technology is transforming everything. AR (Augmented Reality) is providing visual simulation and is significantly augmenting learning. This allows teachers to intensify lessons and textbook content, engage students by transforming the object or place being studied—or even by transforming the classroom itself. AR has the power to supplement education across the curriculum and actively engage learners in a way that is meaningful and aligns with multiple learning styles. Artificial Intelligence (AI) is seen in most accustomed intelligent digital assistants like Google Home, Siri and others for help in everyday life. AI can hop into the study halls—by offering insights into learning, getting the hang of, encouraging correspondence, and supporting educators with reviewing. AI opportune with learning procedure and representative tedious assignments, allowing upgraded learning. E-Learning platforms providing computer-based games and apps for digital learning, continue to evolve. Audiobooks, dictation software, and reading apps are already helping students address visual/auditory challenges. This presents opportunities for teachers to plan lessons and assignments that engage the class in device-based learning, thereby creating a more inclusive environment for all students.

16

Energy Management

"Energy conservation is the foundation for energy independence."
— **Tom Allen**

Rapid technology advancement and industry interconnectivity is creating major shifts in industries landscape. Such interconnectivity can be witnessed from the fusion between software and transport industry—Google Maps, Uber, and all other apps that we use to travel between places. The advances in technology have catalyzed worldwide economic growth and have given ways to achieve energy sustainability using many different energy sources. Besides hydropower, other renewable sources are growing steadily. Solar energy is now used in many industrial applications. Powerwall, a rechargeable Lithium-ion battery by Tesla, stores electricity for household consumption as a backup. This disrupted the energy storage industry under the vision of Tesla, where a person can rely solely on domestically generated solar energy for 24-hour household use. The use of photovoltaic cells has also grown, but it is still as expensive as other conventional sources and very dependable on availability of sunlight.

There lies a great difference in electricity generation from region to region; going from 90% fossil fuels like oil and gas in the Middle East, to over 70% renewables in Latin America—mostly hydropower and biomass; whereas in France about 70% of electricity produced is from nuclear energy. Over 20% of the energy generation worldwide is from renewable sources. Of the 20%, close to 5% is hydropower followed by wind and solar with 5% each, tidal and geothermal power produces less than 1%. Renewables have been a rapid growing sector led by new sources like biofuels. Traditional biomass consumption is being substituted by other renewables. Saudi Aramco, Sinopec, China National Petroleum, ExxonMobil, Royal Dutch Shell, BP, Total, Lukoil and Eni are some of the biggest oil and gas companies in the world.

According to Fortune Business Insights, the global Energy Management System market valued at USD 18.20 billion (2018), is expected to reach USD 48.9 billion by 2026 with a CAGR of 13.7%. The energy industry has gone through major transitions in the past years. About 175 of 2500-megawatt production of coal power plants have been retired in the U.S from 2009. According to SNL energy, same amount of coal power generation plants will retire by 2022. The renewable energy sector has seen decent attraction of funds in the past few years. Governments and private companies are heavily investing in renewable energy research and development programs to generate cost efficient renewable energy solutions. Although giant strides in energy sector are infrequent, changes with the potential to disrupt the industry can take place swiftly. The increasing capabilities and speed of computers are significantly impacting the energy business.

With the help of sensors and real time acquisition, it can be inferred that energy industry has seen an exponential growth in volume, variety and velocity of data gathered from operations. For instance, British Petroleum (BP), in their UK North Sea project used big data analytics to process huge geo-science datasets. Dataset from 500 wells was analyzed in few seconds, whereas a 100 well dataset would normally take a geologist a month to analyze. The critical analyses of large data in few seconds rather than months provide vast opportunities. This has revolutionized the drilling of oil and gas well, and thereby increasing the overall performance of operations. The future of energy is and will be highly data driven, where companies invest in gathering high volumes of data for analyzing, understanding and making insightful decisions.

Automation is already boosting opportunities in the energy industry. In the long run, automation may include deployment of robot in areas where human intervention is impossible. Self-driving vehicles are likely to decrease energy consumption. There has been a rise in traction gained by automation R&D in the recent years.

Vehicle manufacturers such as Mercedes, Tesla, and tech giant Google are leading the innovation in automation. The self-driving cars of Google registered over a million miles on the roads in US. The modern cars have automated mechanism for specific functions too. Automated cars are making their way to market. A large chunk of car market will at least have partially automated vehicles in upcoming years. The off-grid fuel cells industry is also rapidly advancing. Hydrogen is the energy source for fuel cells, and the advancement in the fuel cell technology is very much dependent on development of infrastructure and storage of hydrogen. Fuel cells offer relatively more efficient electricity generation than combustion methods and fewer emissions are generated on extracting hydrogen from low carbon sources. Energy Elephant is an energy management company that makes energy decisions better. The software saves time by automating day to day energy data processing in an organization. It combines all energy data in one location—analyzing, identifying and sharing information. Events2HVAC reduces usage of energy by 20% to 40% returning Heating, Ventilation, and Air Conditioning (HVAC) to unoccupied events. Eniscope, Fabriq OS, Digitalenergy Professional, ePortal are few other alternative energy management software companies for businesses.

The supply and demand of energy has been reshuffled due to the geopolitical change and instability. Despite low prices of oil, gas and coal there has been record investment in clean energy technologies. According to a study by BloombergNEF for the United Nations Environment Program and Frankfurt School's UNEP Center, investment in new renewable energy is on course to total $2.6 trillion through the end of 2019. The investment in solar renewable energy leads with (USD 1.3 trillion), wind (1 trillion), biomass and waste (155.5 billion), small hydro (42.7 billion), biofuels (27.3 billion), geothermal (19.8 billion). *"Investing in renewable energy is investing in a sustainable and profitable future. It is clear that we need to rapidly step up the pace of the global switch to renewables if we are to meet international climate and development goals."*—Andersen, executive director of UNEP. China has seen the most investment in renewable energy capacity with over USD 700 billion, followed by US with over USD

300 billion and Japan, Germany, UK, India, Italy, Brazil, Australia and France under USD 250 billion.[72]

Digital disruption of the energy sector is evidencing new threats too, the cyber-attack. The new-gen technologies like batteries and grid embedded generator are posing threats to cyber security of grid systems by making them more vulnerable to attacks.

Impact of innovation on energy market can be visualized from affected costs and improved quality of energy service. The energy markets in U.S, Europe and around the world are going through dynamic shifts. Energy utilities are always being challenged to keep up governance that plea minimization in greenhouse gas emission, use of inexhaustible energy, enhanced energy storage and smart grid technologies. Utilities are using proactive methods to stay ahead of the possible new entrants in energy market. This trend is evident from the developments in the Duke Energy, the largest investor owned power utility in the US. *"We see a huge opportunity revolving around the Energy Internet of Things. We are tightly focused on providing utilities the tools that need to leverage data from connected equipment and devices to improve grid reliability and resiliency, enhance customer engagement or monetize distributed assets, whether it's renewable power generation, energy storage, smart investors, home appliances and devices."*—John Mclean, former Autogrid's director of product marketing.

Duke Energy invested USD 80 million in 2017 to add new connected devices, networking and decision support software to critical equipment. It introduced 20,000 wireless sensors and ICT in control plants which enabled data streaming over network, empowered employees to identify and react to issues before they evolved into genuine threats.

Google entered the energy business as a leading buyer of green power and capitalist in ingenious energy companies. It achieved 100% renewable energy target. This tech giant's renewable energy purchases exceeded the amount of electricity used in its operations in 2017. Google held contract to purchase 3 gigawatts of output from renewable energy projects, which is by far one of the largest renewable energy purchased by a corporate entity. These contracts led to more than USD 3 billion in new capital investment around the world as of 2018.

Hexcel, a company which uses ultra-lightweight materials for planes, cars and wind turbines, made a breakthrough in carbon fiber and other advanced lightweight composites for aerospace, defense, transportation and wind turbines. This empowered Original Equipment Manufacturers (OEM) to accomplish much higher fuel economy than conceivable with conventional materials like steel. The automobile manufacturer like BMW launched its i3 electric car with 124 MPGe and airplane manufacturers—Airbus and Boeing have incorporated lighter weight materials to increase efficiency. NRG Energy is ahead of most of their competitors in noticing that utilities alter their conventional business model or risk extinction.

Panasonic's Yoshio Ito and Tesla's Elon Musk have devices to supporting Tesla's revolutionary "Gigafactory"—a multibillion battery production line projected to break ground in the US. Alongside the rise of an enormous lithium-ion battery, factories gave scope for stationary batteries, which mingled with Panasonic's solar PV cells. This set a record for sunlight transformation efficiency.

The pre-eminent turbine supplier Vestas is behind numerous advancements in wind turbine innovation that set wind power as the source of electric generation with least cost across many parts of the world. Vestas has sector's largest machine to date. On an average, more than 7000 homes are powered by the megawatt turbine designed for seaward wind ranch.

Swasea based Specific, an innovation and knowledge center has developed active buildings that generate, store and release their own solar energy. PivotPower have rolled out 20 ~50MW batteries located at grid substations, that provides buffer storage for rapid charging electric vehicles. Equinor's Hywind floating wind turbines projected that the floating wind can be a viable option and intends to build a full scale 380MW energy unit.[73]

Siemens AG, a global technology powerhouse with excellent engineering, innovation reliability and quality for over 150 years. It is the world's largest producer of resource saving and energy efficient technologies in wind turbine construction, power transmission and automation solutions. Cypress Creek—developed, built and operated solar facilities across the US, producing over 3 gigawatts of solar developed energy. NextEra Energy, serving over 460,000 customers, is the leader in battery storage and world's largest generator of renewable energy from sun and wind. Vestas, GE Energy, Ostred, Suzlon, Berkshire Hathaway, Avangrid, EDF, SEAS-NVE, Georgia Power, Borrego Solar Systems, Duke, Xcel, Geronimo, AEP, Ecoplexus, Invenergy are some of the key renewable energy companies.

According to the International Energy Agency (IEA), 5.6 billion air conditioners will be built globally by 2050. Traditional HVAC systems are often polluting as they use HFCs (hydrofluorocarbons) which produces 3,000 times stronger greenhouse gasses than carbon dioxide. New heating technologies that are smarter and much more efficient are leading the way to improve air quality.

Organizations such as Bezos, NEA, and Goldman are investing USD 4 trillion in Fusion Energy to disrupt the electricity market. Fusion energy is produced by fusing light atoms at extremely high pressures and temperatures together, such as hydrogen and its isotopes (protium, deuterium, and tritium). Plasma becomes the fourth state of matter at high temperatures encountered by any substance (the other three are solid, liquid, and gaseous). Plasma is an electrically charged fuel which is then used to generate electricity. Fusion energy generates abundant clean energy and also helps in meeting global electricity demands. The concept of fusion energy has always been convincing, as it is an unlimited energy source that is safe to operate and generates zero carbon emissions.

The pace of recent developments in energy sector is a clear indication that the sector is going through major disruptions. The present technologies and processes will be replaced by much more advanced technologies. The same trend can be traced from the history when nuclear was considered as a technology to rely on, but was replaced by conventional power generation methodologies like combined cycle gas turbines in 90s. The point here is that these sudden technological shifts are now intensified by pervasive digital innovation. This digital disruption will force energy businesses to adapt and change more and more frequently in the future.

17

Digital Public Citizen Services

"There is no greater challenge and there is no greater honor than to be in public service."
— **Condoleezza Rice**

The availability of information at lesser or no cost and with easy access is the hope of the government. These expectations from citizens and businesses can only be met by a successful digital transformation of public sector. A survey by McKinsey showed that capturing the full potential of government digitization could free up an economic value of USD 1 trillion annually worldwide, through improved cost and operational performance. The efficiency of the system is enhanced by integration, collaboration, shared services, improved scam management and advancement in productivity. Governments at any level, national, state or regional cannot afford to miss on these savings.

Governments around the world, have recognized the importance of digitalization and have included digital transformation as one of their primary agenda. Online services are provided by around 130 countries. In Estonia about 1.3 million residents are able to vote, pay taxes and access several online services using their electronic identification cards. The consolidation of many data sources regarding the government into a sole platform by the social aid information system of Turkey, enables citizens a quicker access on various aid programs.

The government at all levels are at the edge of transformation, as the digital operating models are counter parting analog models. The services delivered by the government depends on the increase of tech experience generations, budget shortcomings and inflated entitlement expenditure.

In order to get a clear picture of digital transformation in public sector 'Deloitte Digital', a multinational professional services network overlooked at more than 70 countries, interviewed around 1200 officials, 140 government leaders and external experts. The research focused on issues such as user focus, workforce skills, strategy, procurement and culture, as these were the areas considered to accelerate digital transformation by public leaders. In spite of all the attempts and advancements made by the government towards digitalization, most of them are still far away from capturing the benefits. To successfully utilize the benefits of digital transformation, governments should move beyond just online services. A deeper transformation that can improve productivity, end-to-end collaboration and scale resources with process efficiency is vital.

It is difficult to invest and produce enough economies and benefits, on range of platforms, when the data and systems are held by many departments, functions, differing taxonomies and access credentials. IT infrastructure and the common components of a business makes it hard to connect internal systems—be it a government worker, a business user or another intergovernmental office, creating a seamless user experience for the end user. Complications arise in projects such as government transformation, where functional skills and proficiency are available only at a very high cost and are often in less supply. A clear technical strategy and leadership poised to drive the transformation is what separates digital leaders from the rest.

According to a study conducted by IDC with Salesforce sponsorship, about 80% of the US adults used smartphone against 45% of the government employees. The citizens are increasingly adopting and becoming aware of digital technologies in their day to day lives, thereby compelling government agencies to align to this trend. "The government of the future will be organized around the citizen, and no longer will the public have to navigate a complicated bureaucracy to figure out where to get service. Through modernizing technology, government organizations will not only add functionality but also gain the trust of their customers and constituents." Casey Coleman, Salesforce's Senior Vice President of Global Government Solutions.

Department of Motor Vehicles (DMV) is using online tool, enabling citizens to schedule appointments, renew vehicle and driver licenses etc. New York city's electronic water meter collects data on usage of water. The implication of digital technology has almost cut off human intervention, where it required people to physically go around and read meters.

"Transformation means more than fixing websites. It goes deeper than that, right into the organizations behind the websites. There is no logic to it: Digital service design means designing the whole service, not just the digital bits"—Chief digital and chief data officer for the government of UK. To remain competitive among economies globally, digital and information excellence is essential.

In a global scenario the digital transformations in government isn't restricted to just economic shifts. Associations like OCED (Organizations for Economic Co-operation and Development) and EU (European Union) invite member states with directions and proposals to bring their governments closer to citizens. Like, how in digital transformation, 'customer experience' is of topmost importance for private sectors; 'citizen experience' is of topmost importance for public sectors.

Despite the challenges that public sectors face in incorporating digital transformation, there are number of examples illustrating successful government initiatives. The website "GOV.UK" of United Kingdom government, serves as an information hub for rural population to get access to all the government programs. The success of GOV.UK is enabling government to provide services to its citizens, businesses and others under the government. The capable leadership from various public departments led to the success, which in turn led to the need of expertise.

Training or reskilling of workforce is a mandate to incorporate any new technology or new workflow in any organization. Re-Skilling programs like the 'six-week digital boot camps' run by the UK's Department of Work and Pensions is an example, driving successful digital transformation. Organizations require asserting the humanitarian value of public service, diminishing impression of hindered bureaucratic darkness, to draw in more youth employees with more of private segment opportunities. "You could recruit people into a bad environment. Those people would tell their friends, and things would fall apart after that. It is not enough to just recruit."—Greg Godbout, Chief Digital Officer, TechFlow Inc.

Digitalization has made data security, a topmost national security concern. Cyber-attacks and critical systems failure have been identified as two of the most dangerous global risks by the World Economic Forum. For government, cyber-attacks not only cause financial loses, they also pose serious reputation risks.

Higher wages, a more open culture, clearly defined mission and vision of private sector often makes it challenging for government organizations to get right talent. To attract the IT talent, governments have found many ways—for instance, a major part of government IT infrastructure in South Korea is focused in a few information centers giving a number of e-government facilities to public. In UK, government is offering fast track career opportunities for high performers to attract talent from private sector. The lead of digital services and government CIO spend major part of their careers in private segment.

Governments can utilize data analytics and big data to improve decision making. In fact, use of data to make informed decisions has already been adapted by governments around the world. The government of U.S. is one among the most dynamic in influencing the data analytics to support "choice making." They made open data legal and privacy framework in 2009, which led to creation of data.gov, a repository of government tools, resources, and information on anything from energy and science to global development and health.

In the US, businesses and professionals get help in conducting research, developing web and mobile apps from more than 85,000 data sets online. To populate these data sets, government departments have contributed their most valuable data and best practices. The government of US has arranged numerous competitions like Apps for Democracy and Apps for America to draw skilled designers to develop applications that utilize government data, as the citizens prefer tools that are adaptable and user friendly.

Most developed and major economies have created a national cyber security strategy that involves information sharing mechanisms, data breach security and many advanced data security tools that can detect and prevent data thefts. The UK Fusion Cell serves as a great example, where masters from private and government sector are brought together in threat analysis and information sharing hub. The primary duties of the governments are to detect fraud and crime for the betterment of its citizens. Performance in these areas can be accelerated by ambient computing. Greater business value can be achieved when sensors are used with data integration, data analytics platforms and a sound cyber security system.

The public sector CIOs are choosing from the large systems that supported their agencies for decades and the digital wave of social media, mobility, analytics and cloud (SMAC). In an intention to advance their traditional transactional systems, these CIOs are considering the ways they can utilize the lessons learned over time to motivate modern administrations and developments at the core. This "major renaissance" is required and still requires a lot of integration, coordination, and adjustment by IT leaders of public sector.

Governments around the world have led, adopting technologies like artificial intelligence in critical areas like national intelligence and defense. However, they are lagging behind their commercial counterparts in using AI and language processing capabilities to improve operational efficiencies. The efforts and intelligence of public sector can be augmented and amplified with machine learning, generating data driven insights that will improve decision making in a variety of ways.

The experienced public sector workers are quitting while the agencies with modern proficiency youth workers are replacing them. This varying platform is pushing innovation leaders to re-evaluate their reach to skill management for present and the future. The modern public sector leaders are centering their focus on the workforce which will supply value from these investments. Employing the desired capability with right talent sets will result in a combination of overcoming long simmering challenges and leveraging new opportunities.

Digital technologies have raised expectations of customers for getting their needs delivered quickly and in a frictionless manner. However, a different experience is offered by regional governments like—forms have to be printed, then mailed, payments taken only in cash or check, appointments that have to be conducted in person and in offices open only from 9AM to 5PM. This trend has created a scenario where people find it easy to compare restaurants, pubs than public schools and organizations. It's a missed opportunity for governments. Public expects government to progress comfort of public living, and also to illustrate their esteem and raise civic engagement.

It may be, starting a new business or a driver license renewal, some local governments are utilizing innovation and client centered mentality to innovate and serve the public in a better way. In order to have a significant effect, government administrations are scaling-up its efficiency by merging and collaborating online and offline capabilities while focusing to construct a digital audience. The government is providing customer service touch points, subscription opportunities and community events on their social media and websites.

Smart segmenting is required to serve large populations, to achieve the private sector standards of services and engagement. Government organizations must also segment their customers while keeping the citizen's privacy as topmost concern. To attract and engage citizens in programs and initiatives of regional government, tailored communications are of prime importance. For example, for efficient sanitation services, junk pick up note is offered by the city of Louisville through GovDelivey, sectioned by area. Over 12,000 families have subscribed to get alert message of the schedule of junk pick up a week before and one day prior of the schedule.

The governments are capable and should consider that the right technology with right associations will enable radical impacts. The governments worldwide are on way to provide an actual public associated experience. Without the collaboration of better communications methodology and right innovation, public citizens cannot get a clean and a persistent experience from the public sector. For instance, in Canada, digital government provides a forum to discuss best practices, collaborate on projects and team up on ventures with different executive levels. Collaboration is critical to success—regardless of whether it is through gaining from international expertise or ensuring that departments are internally dedicated to, and prepared for implementing new technology. Digital government acknowledges improving service delivery to citizens that starts with digital tools, but ultimately requires leadership far beyond IT. It is important to benchmark the key challenges faced by government at all dimensions and build a comprehensive report that includes combination of surveys and interviews along with expert opinions.

Reaching customer expectations is imperative, and government's customers both external and internal assume that services ought to be accessible whenever required, through whatever source or channels. Ensuring responsive and multi-channel access to services are all critical parts of vision for the present-day government. To meet these expectations, organizations must assemble tools, talent and knowledge of client (Public Citizen) needs that aids to convey insightful and efficient digital services. Government and industry experts are now exploring how businesses can meet these demands and transform their relationship with digital citizen.

18

Women Entrepreneurship

"I want to empower and inspire women to fearlessly live their dreams!"
— **Ankita Vashistha**

Women are extremely focused, determined and have a proven history of success. They have the magic of wonderful innate qualities and characteristics that they bring to the forefront everyday as individuals, mothers, sisters, wives and successful business leaders. Women face extraordinary challenges in the multiple roles they play and have the added burden of balancing work life. However, they have developed expertise and knack of tackling situations and overcoming them very well. Women today have broken the glass ceiling in the biggest of corporations and are emerging as exemplary innovators and entrepreneurs. Their contribution is redefining technology, engineering, fashion, agriculture and a myriad of industries.

Women Entrepreneurs are playing a vital role in the economic development of their respective communities and countries. Women have greater emotional intelligence, which makes them the natural choice for leadership roles. Women need to think big and define their goals. Employers and employees need to be aware of the incentives and support available to working women, as its important to retain a critical segment of the workforce. It is imperative that we create inclusive initiatives and strategies so that women can be part of the work force. The benefits of understanding these unique motivations are imperative to retaining women's experience, intelligence, empathy and drive, all of which helps fuel our economy.

Only 10% of startups are represented by women founders and of this the ones that get funded are even less. A recent report by FT, showed that out of almost 2,000 founders who received venture funding in the UK, only 9% were women. The number are pretty similar or worse in other parts of the world. In terms of board seats, while women hold 22–23% of these in companies with market caps above $50B, that only represents 7% of NYSE companies and 1.5% of NASDAQ companies. In the masses of smaller revenue companies' women hold between 9% and 15% of board seats (data from The Business Journal 2018 research).

Working mothers are a very important segment that needs to be empowered and, it's critical that leadership and organizational policies respect their need for flexibility. A majority of women (75%) ranked their manager's and organization's support for having the flexibility to balance work and professional life as the most critical criteria in job selection in terms of feeling respected at work, part of a study released by the "Mom Project." Written policies and HR workshops are not enough. Leadership needs to visibly support efforts to create work-life balance and demonstrate a genuine understanding and respect for the need for flexibility for their employees. Women report that feeling respected in the workplace and understood for their need to balance between family and work has a significant impact on their overall workplace satisfaction, performance and desire to stay at an organization.

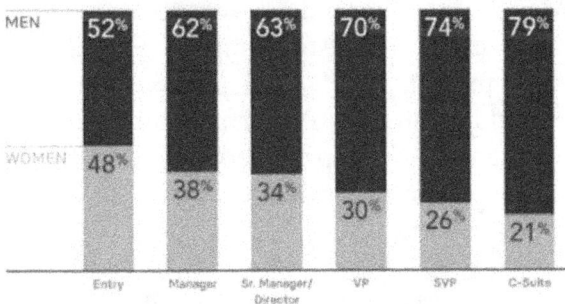

Figure 2: Gender Representation in the Corporate Pipeline

Over the past five years, the number of women in senior leadership has grown. Still, women continue to be underrepresented at every level.

*Source: Leanin.org; Women in the workplace 2019 report

In the technical field in certain countries, there is almost a 50-50 balance in the workforce. The presence of women in leadership roles has also increased significantly and is over 30% in most growth markets. Women are doing their part. For more than 30 years, they've been earning more undergraduate degrees than men. They're asking for equal pay and career promotions at the same rates as men. Over the past five years, the number of women in senior leadership roles has grown, but women representation is still low at the C-Level and board executive level, as reported by the Leanin organization. The BNP Paribas Global Entrepreneur Report found that the success rate of women entrepreneur has significantly increased over prior generations. Education and healthcare industry are predominantly administered by women entrepreneurs.

The leading women entrepreneurs globally include amazing leaders from biotech, fashion brands, media, celebrated authors, venture capital and others.

Kiran Mazumdar-Shaw is an Indian billionaire and India's first biotech entrepreneur. She founded Biocon in 1978 at the age of 25. Biocon is the only second company from India to list a USD 1 billion IPO on its first day of trading. Biocon is India's largest producer and exporter of enzymes. In 2014, Dr. Kiran was awarded the Othmer Gold Medal, for outstanding contributions to the progress of science and chemistry. According to Forbes, 2019 billionaires net worth projects USD 3.5 billion. She held 60th position in 'The World's 100 Most Powerful Women 2018'. In September 2019, Biocon announced a price reduction of insulin to 10 cents per day vs the current price of 5 dollars a day. An amazing innovation and a boon to mankind.

Tory Burch is an American fashion designer and businesswoman. She is the Chairman, CEO and designer of her own brand, Tory Burch LLC which is an American fashion Label. She is the billionaire queen of USD 200 ballet flat. As an art history graduate, she worked for designers like Ralph Lauren and Vera Wang before starting her own brand in 2004 and was endorsed by Oprah Winfrey in later years. A separate active wear line, Tory Sport was launched in 2015. In 2015, she established the Tory Burch foundation which intends to empower women entrepreneurs through low cost loans and mentorship programs. She has a net worth of USD 850 million, and holds 29th position in 'America's Self-made Women 2019' by Forbes.

Sara Blakely is an American billionaire businesswoman. She is the founder of Spanx. Spanx Inc. is an American underwear maker focusing on shaping briefs and leggings. She started her business at the age of 29 with her lifetime savings of USD 5000 as an attempt to come up with something that can be worn under bleached slacks. Her career took a turn 6 months later when a one-time Disney world greeter and door to door fax machine salesperson found that her brand of shaping underwear was one of Oprah Winfrey's favorite things. Since then her business has grown from one hit wonder, to a worth of USD 250 million in annual revenues with net margin profits estimated at 20%. Today, Spanx sells its undergarments, leggings, and maternity wear in 65 countries. She has a net worth of USD 1 billion and holds 23rd position in Forbes 'Real Time Net Worth' list.

JK Rowling is a British novelist. She is known for writing The Harry Potter Series. Rowling was working for Amnesty International as researcher and bilingual secretary. She got the idea of Harry Potter series in 1990, while on a delayed train from Manchester to London. Rowling was a single mother and was broke. The first novel in the Harry Potter series got her success. She penned 6 sequels and sold 500 Million copies worldwide. The Harry potter series was also adapted to very popular series of films. The film franchise and theme park also kept the profits coming in. She is the world's first billionaire author.

Oprah Winfrey is one of the most influential women in the world. It all started with her talk show, The Oprah Winfrey Show which is the highest rated program of its kind. Her interview with Michael Jackson in 1993 is the most watched interview ever with an audience of 36.5 million. She was born in rural poverty and was brought up by her single mother who depended on government welfare payments in a poor urban neighborhood. Winfrey became a millionaire at the age of 32. She built her business empire from her talk show. In 2011, after bidding goodbye to her daytime TV show, she launched cable channel OWN (Oprah Winfrey Network) partnering with discovery. In 2017, Discovery Communications purchased 24.5% of OWN from Oprah. Forbes estimates that Oprah's share of OWN is worth about USD 75 million. Her shares in Weight Watchers have grown from USD 43.5 million to USD 400 million, of which she has been the brand ambassador and board member since 2015. At present, she has a net worth of USD 2.7 billion. She was the richest African American of 20th century and North America's first black billionaire.

Kylie Jenner, Founder of Kylie Cosmetics. A new age entrepreneur and youngest self-made billionaire, who built out her billion-dollar beauty brand solely using online marketing and social media. Amassing 153M followers on Instagram, Kylie Jenner is one of the most followed people on the popular photo sharing platform, Instagram. In November 2019, Kylie Jenner sold her majority stake of 51% in the company to leading global brand Coty for $600M. Kylie Jenner is an example of a completely online digital entrepreneur riding on latest trends and social media flurry to help her rise in the beauty world.

Ankita Vashistha is the founder of Saha fund. India's first Venture capital fund for women entrepreneurs. She had this idea as an investor, when she noticed that women never made pitches during the meeting even though they are cofounders of the startups. Also, there were only a handful of women investors and almost no women focused investor. The focus on diversity had to consciously be made and aware of with an investment vehicle and platform like Saha Fund. The fund's investors include leaders from the industry and investors looking to empower and support the power of women entrepreneurship, empowerment and engagement. Her mentors and investors include leaders like Biocon chairperson, Dr. Kiran Mazumdar-Shaw, Zia Mody, and Avinash Vashistha, former Accenture India chairman. The fund looks to invest and work with portfolio companies that embrace and understand the value of diversity and the positive impact it brings. After all, a culture that values diversity makes a stronger business and will ultimately make the economy and those part of it stronger and more successful.

"If your actions create a legacy that inspires others to dream more, learn more, do more and become more, then, you are an excellent leader."
— **Dolly Parton**

"We need to reshape our own perception of how we view ourselves. We have to step up as women and take the lead."
— **Beyoncé**

"A woman is like a tea bag – you never know how strong she is until she gets in hot water."
— **Eleanor Roosevelt**

"Destiny is a name often given in retrospect to choices that had dramatic consequences."
— **J K Rowling**

19

Top Global Innovation Leaders

"Some people see innovation as change, but we have never really seen it like that. It's making things better."
— **Tim Cook**

The last decade has been clearly a decade of Innovation. Many innovators from across the globe have changed our lives and reshaped industries forever.

Amazon – To anyone who's been held-up in a grocery checkout line for 20 minutes because of the person in front of you is fighting over an expired coupon; here is the solution—Amazon Go, a physical food/grocery store with the best feature: No checkout lines. Using the company's "Just Walk Out" technology, customers use an app to enter the store, pick out the products they need, and then head out. The amount for the groceries is automatically deducted from customers' Amazon account.

Amazon offers a paid subscription service called Amazon Prime, on monthly and yearly terms. With Prime, customers avail free delivery, that too within a couple of hours from order placement. It also has additional benefits like streaming ad-free videos and music. Amazon has over 100 million Prime subscribers worldwide. Prime Air is a deal-breaker—that allows customers to get their orders delivered in 30 minutes or less, via drones, which drop them at the doorstep. That's right, flying robots will deliver packages immediately, as soon as an order has been placed.

Alexa or Amazon Alexa is a virtual assistant developed by Amazon. It was first used in Amazon's brand of smart speakers Echo and Echo Dot. Alexa is used in voice interactions, to get real-time information such as news, sports, and weather. It can also be used to control home automation systems.

Amazon Web Services, another service from Amazon which provides on-demand cloud computing platforms for customers, via paid subscription services. They provide various services like analytics, networking, storage, networks, and the Internet of Things.

Apple – Apple iPhone stormed the smartphone industry in 2007. Apple has sold over 1.2 billion iPhones worldwide. iTunes churned the traditional music delivery model overnight. It plays tracks from CD with the help of smart playlists, and disrupted the conventional music sharing model, making CDs outdated. iTunes demonstrated that computerized media dispersion is possible on a substantial scale, bringing entertainment to the digital age. Sound optimizations and wireless transfer of iTunes library are dashing features of iTunes. It became an integral part of Apple's best-selling product. Apple Watch is another product from Apple, launched in 2015. It is a smartphone watch that includes sensors for fitness tracking, monitoring, and reporting. Apple Watch can track your walk, run, sleep, monitor heartbeat, and control a smart home.

Microsoft – Microsoft Skype Translator is an online translator that assists users to communicate in over 10 languages on voice calls, and over 60 languages on instant messaging. Video to language utilizes Recurrent Neural Networks (RNNs) to describe video content in the preferred language. Microsoft Cognitive Services allows developers to develop smart feature apps by implementing machine learning APIs. The features consist of vision, voice, facial recognition and language understanding.

Google – Google launched Google Analytics, after acquiring software from Urchin Corp in 2005. It is a freemium web analytics platform that analyzes and reports web data for usage optimization.

Android is a mobile OS (operating system), that utilizes a modified version of Linux Kernel, and other open-source software.

It was designed for touch screen smartphones and tablets and is now the highest-selling OS in the market, with over 2 billion monthly active users. Google's virtual assistant utilizes artificial intelligence (AI), which is a two-way interaction platform based on voice recognition. It was initially launched as Google's messaging app, Allo.

Facebook – In 2017, Facebook announced its collaboration with Apple TV, Amazon Fire TV, and Samsung Smart TV, to develop Facebook TV app. The app allowed users to watch and share trending videos worldwide. Facebook Messenger Chabot employs artificial intelligence. Facebook has partnered with numerous brands that allow customers to receive update alerts. Facebook Chabot's are live round the clock and it sends instant responses to user inquiries. Although Chabot's sole purpose is to address FAQ, it still contributes its valuable time to the customer service workers. Facebook Chabots are great for e-commerce business. It can retain tons of data from customer conversations. This data is then put into action. It further reaches out to targeted social media and omni channel marketing targets. Facebook Analytics is excellent for visualizing the activity of messenger bot. One can see the total users that are engaging with the bot, the demographics of these users, and the retention rate. With e-commerce software capabilities, the Facebook chabot can identify the item which the customer is looking for and connects them to that item via social media call-to-action (CTA) buttons in a matter of seconds. Facebook Messenger, with 1.3 billion active users worldwide, is ranked second among all the mobile chat apps. Facebook, however, remains the most popular social media network. With a 73% satisfaction rate, live chat software has emerged as the top way for customers to interact with businesses.

Uber – Uber launched its online food ordering and delivery platform in 2014, called Uber Eats. It let customers view the menu, order, and pay for food to eateries around them. Uber customer experience—Uber locates and detects 'device and sensor level information' by tapping into device accelerometer to recognize the frequency of braking, device movement, and overall speed of the vehicle. Uber analyzes the data gathered through the app and enables drivers to receive daily reports. It gives them an idea of how effective they were in guaranteeing customer satisfaction. Feedback on driving patterns are compared with other drivers in their city and suggestions are given for a smoother and healthier trip that enhances the passenger ride experience. Uber presents an unparalleled opportunity to improve road safety in new and innovative ways—before, during and after every trip.

'Uber Health' is a HIPPA compliant technology. It is a platform where doctors can arrange pickup rides for their patients. The main aim of this specifically designed service is to ensure safe transportation and be patient-friendly. There is a unique dashboard designed for scheduling rides for patients. The healthcare associate schedules future patient appointments. Uber drivers contact patients via text or call, collecting detailed information on their pickup and drop location and timings. The driver reaches the pick-up location, picks up the patient and drops, based on the schedule. Health associate doesn't require phone calls or follow-up regarding a patient's safe journey.

Boeing – AnalytX uses its aerospace expertise with data-based information to optimize the operation and the mission. The enabled product and services of Boeing AnalytX are offered in three categories to the customers—Digital Solutions: the set of analytics-enabled software applications addressing the needs of the crew and fleet scheduling, flight/mission planning, operations, maintenance planning/management, inventory, and logistics. Analytics Consulting: the services include a group of aviation, business, and analytics professionals who help customers to improve their operational performance, efficiency, and economy. Self-Service Analytics: it opens up the data behind the digital solutions for customers to explore and discover new insights and opportunities using Boeing analytics tools.

General Electrics (GE) – The Digital Wind Farm is a comprehensive hardware and software solution, comprised of customizable 2-megawatt and 3-megawatt wind turbine products and a suite of applications. It is built on the predix software platform to bring new value to wind farms. This wind project, has the latest hardware, including an assortment of rotor diameters, tower heights, and turbine ratings. Modular turbine technology enables one to vary turbine parameters to meet specific conditions and economics. 'Centricit's solution 360' is a GE predictive tool that helps hospital administrators to streamline clinical collaboration with unaffiliated clinicians and patients. It reduces duplicate imaging so that it avoids unnecessary patient transfers, lower CD (Compact Disk) distribution costs, and enhances referral relationships—all with no capital investment. The Centricity Cardio Enterprise solution is an integrated cardiovascular clinical informatics system. It includes both CVIS (Cardiovascular Information System) and CVPACS (Cardiovascular Picture Archiving and Communication System) functionality to provide clinicians full access to a single, comprehensive cardiovascular patient record. Centricity Cardio Enterprise, features web-enabled technology, capable of driving workflow efficiencies and integrating with current IT systems.

Tesla – Tesla made a breakthrough in the electric car manufacturing segment with its long battery range and cutting-edge software technology. Tesla designs cars by combining both the hardware and its software. With application updates, users can treat their Tesla cars as an App. The cars come in with an Autopilot mode, which provides driver assistance that is capable of braking, steering, and throttling in particular scenarios. It is also capable of changing lanes by using turn signals.

IBM – IBM Medtronic built a cognitive mobile personal assistant app to assist with daily diabetes management. It uses IBM steams and Watson's platform for health, to help improve the lives of people with diabetes. The app provides actionable glucose insights and predictions. The IBM operational decision management software was used by "Brownells" and "Salient Process," in order to optimize the management of business rules and process regulations which could affect the acquisition experience through the company's rules engine. IBM's latest innovation is an AI-powered robot microscope that can help clean up water resources. The AI-powered autonomous robotic camera developed by IBM has the power to monitor the behavior of microscopic plankton that helps identify chemical pollution levels and temperature variations of water. The data from these cameras can give ideas on the factors affecting the quality of water resources and life in it.

DARPA – Defense Advanced Research Project Agency (DARPA) is a defense agency of the United States. DARPA develops technology for the US military. Some of their technology developed for the US military has also influenced technologies in many non-military fields. Innovations from DARPA has led Internet development. 'Global Positioning Satellites' and Automated Aerial Vehicles/drones' are widely used discoveries of DARPA. 'Memex' is a DARPA program used for context indexing and web searching over the internet, helping in countering human trafficking.

SAP – SAP is an amazing enterprise driven by innovation. In 1972, SAP founders pioneered standard software for financial systems to run customers' business processes in real time. SAP, since then has become the leading enterprise resource management platform, running payroll, human resources, client relationships, supply chain and manufacturing etc. SAP has revolutionized data management with SAP HANA, and put customer experience at the core of operations. SAP is a great example of how a leading enterprise software company is Future-proofing their customers' business and building systems that provide a foundation for growth and innovation. With AI and other new technologies, client experiences need to be hyper-personalized and outcome-based. Like some of the top innovators recognized in this chapter, SAP has designed and implemented an economic innovation ecosystem which brings and connects together, the innovative power of top-tier academia, emerging technologies and open innovation to re-imagine client experiences. This is driving large scale adoption and fundamental changes for various industries.

20

Delivering Innovation
at Scale

"Innovation is the change that unlocks new value."
— **Jamie Notter**

Digital is not anymore, a mere technology, it is now part of everyday life – live, dine, exercise, bank, shop, learn, travel, work etc. Enterprise Innovation at Scale is about delivering a re-imagined and exemplary consumer/client experience, through innovation, artificial intelligence, machine learning and intelligent automation. Experiences are being redesigned to wow consumers across retail, media, banking, restaurants, food delivery and even supply chain. Each element of an enterprises business and consumer is being re-imagined to deliver, the wow unexpected experience.

How can enterprises deliver innovation at scale? In simple terms, let's think of three key elements to effect digital transformation – Re-imaging Consumer Experience, Bring in Innovation and Implementing through Intelligent Automation. The traditional legacy businesses are being transformed through Robotic Process Automation (RPA), Machine Learning, Artificial Intelligence, Augmented/Virtual Reality, IoT and many other digital technologies. For instance, Disney–All through its long history, it has been an inventive company. However, it has faced many challenges on its way. Recent adaptation of digital technology across its entire business has paid marvelous dividends. Providing customized MagicBand for every visitor at its theme park and hotels is one of the ways how digital technology is used. This has profited both, the company and customers. The quality of experience and efficiency of gadgets have been greatly increased through personal and interactive tracking and sophisticated real time data analytics. Labor scheduling accuracy has improved by 20% in its theme parks, representing an incredible payoff. Elsewhere in its empire, store sales are up by 20% following re-design and embedding of digital technology. Summing up, Disney's approach to technology is transforming customer experience, cross channel interaction and network connectivity.

Innovation at scale for enterprises is a digital transformation process/function which bring three key elements to deliver sustainable impact:

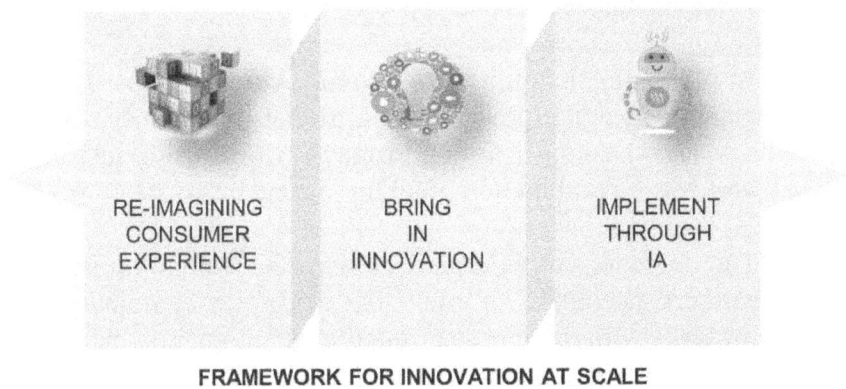

RE-IMAGINING   BRING   IMPLEMENT
CONSUMER   IN   THROUGH
EXPERIENCE   INNOVATION   IA

FRAMEWORK FOR INNOVATION AT SCALE

Figure 3: Framework for Innovation at Scale

Re-image Consumer Experience: Enterprises are working with creative design studios to re-imagine consumer experience through a process of "dreaming." The idea is to take clients and consumers to a state of levitation that can enable them to dream the impossible. It is important to ensure that the dreamers are free from the shackles of technology, practicality and implementation realities. This creates a sweet but near impossible dream. How do we make the impossible, possible!

Bring in Innovation: The world is full of innovators, who have re-imagined, re-designed and implemented amazing innovative solutions to deliver snippets of near impossible dreams. It is possible to explore, discover and bring in these innovations to make the impossible dreams, near possible. The solution design through innovation in most cases delivers the wow consumer/client experience.

The innovative solution includes data and intelligent automation strategy, cloud, mobility, blockchain, AI, social media, IoT, machine learning and cybersecurity with advanced analytics to unlock new intelligence. The solutioning augments the power of humans with artificial intelligence, leverage AI and advanced analytics algorithms to sense, comprehend, act and learn across value chain at an unprecedented speed and scale.

Implement through Intelligent Automation: Todays implementation is agile, powered by artificial intelligence, digital technology and machine learning. Today's workforce consists of digital workers. A combination of all these elements is what we call intelligent automation.

The operations at workplace is set to have both human and digital workers, becoming an intelligent workforce completing the task/set of tasks with the power of intelligent automation, analytics and decision-making capabilities while increasing the productivity and quality and cutting down the cost. In the process of intelligent workforce operations, cognitive technologies play an important role, learning from the best.

The above model of enterprise innovation at scale has been tested across multiple large digital transformation engagements across many of the fortune 500 and global 2000 companies by us. We call this the IaS Model of Innovation. Let's discuss the model and how it was applied to one of the top 5 banks in the US and led by one of the big four firms.

A case study of KYC/AML refresh for US consumer banking and wholesale global banking.

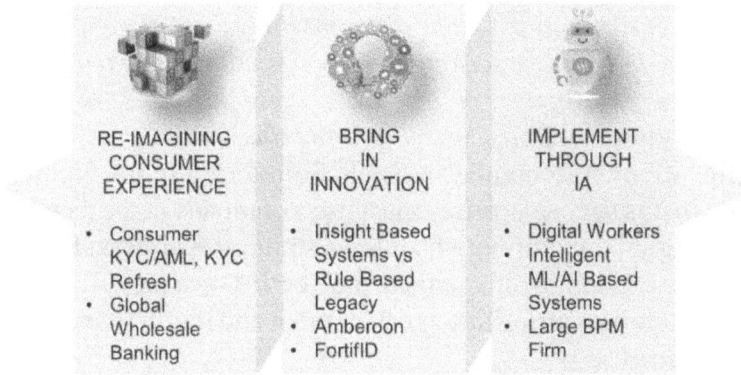

RE-IMAGINING CONSUMER EXPERIENCE	BRING IN INNOVATION	IMPLEMENT THROUGH IA
• Consumer KYC/AML, KYC Refresh • Global Wholesale Banking	• Insight Based Systems vs Rule Based Legacy • Amberoon • FortifID	• Digital Workers • Intelligent ML/AI Based Systems • Large BPM Firm

Figure 4: Framework for Innovation at Scale for Financial Regulation

The key product developed at Tholons Digital is a fully digital insight based, AI/ML fintech platform for Know Your Client/Anti Money Laundering (KYC/AML). This fintech solution incorporated AML data analytics from Amberoon and intelligent automation. This was first beta created for a top 5 US bank with compliance audit review by one of the top big fours. This solution addressed the entire KYC refresh of US consumers and Global wholesale banking. This solution enabled the bank to cover mid-risk clients that otherwise would have required them to quadruple their KYC team from 1200 to over 4000 bankers at a cost of USD 200 million/year. The Tholons Digital fintech solution enabled the bank to deploy at an average cost of only USD 100 million/year and with less than 400 bankers.

This is an amazing validation of the model as it demonstrates that clients are willing to adopt a collaborative best of breed solution, consisting of multiple moving parts and stakeholders. Innovation from smaller startups are being welcomed. Solutions that have not been implemented and proven before are being embraced at scale. Five years ago, no one would have ever imagined that a heavily regulated large bank would ever risk evolving to such a solution so fast. Survival, growth and leadership requires risk taking, grit, the conviction and hunger for innovation.

18.1 Transforming into Digital Business

"Digital transformation means different things to different business. Whether businesses are seeking to optimize manufacturing, improve an in-store experience or create a new business model, the transformation required crosses multiple lines of business."—Laura Merling, VP of Autonomous Vehicle Solutions at Ford. In order to succeed in this age, enterprises must use a company-wide framework for their digital transformation and IoT strategy. This transformation will require alignment across the entire business—including investment dollars, priorities, measurement and metrics, and the path to execution.

Innovation catalyzing digital transformation can be an amazing thing. Companies like Tholons have adapted a unique way to get its innovation core up and running. They believe that the weight of innovation stands on 5 different pillars. These are Clients, Startups, Mentors, Platform builders and Funding. All of these aspects need to work together, in order to run a successful digitally innovative business.

If there is a new solution, that needs to be implemented on a large platform, think about the clients first. Will it add value to use the new solution? If yes, it requires to either develop the solution in-house or avail a more economic option outside. Startups are a great resource to develop niche market products and can add great value. Then comes the mentor, without proper guidance, a plan can go off track and cost the company. The solutions created by the startups are then integrated by the platform builders, followed by the appropriate funds to execute the plan and endeavor.

"Everybody wants to beat the competition; so, product development is often a race to give customers more,"–Forbes. One-upmanship is the bread and butter of the corporate world. In the digital world, there are no laws of physics to constrain innovation. So, adding new solutions can be easy. The focus needs to lie on how the customer feels while using that solution. For instance, 'StubHub' is the world's largest ticket marketplace, with millions of unique visitors and ticket sales every year. The company was struggling to meet the digital demands of processing thousands of transactions every day. To keep up consistency of providing better service to their

users, StubHub created its own private data cloud, with incredible scaling capacities to overcome the problem. This makes sure that company can cope up with the variable level of demands the website experiences on each day. Addition to this the company also has planned to use public cloud to process local transactions in many countries and add features to make selling tickets easier and available for mobile users, and easier for group of people to buy tickets.

Digital transformation creates new revenue streams and improves processes' efficiency, by a series of small and large-scale projects utilizing/installing digital capabilities. In its entirety—it is big. Everything and everyone will not be aligned perfectly, however, sharing an understanding of the goals and providing transparency is paramount. By providing a mechanism for change, it gives businesses an opportunity to succeed.

18.2   People   Primacy   in   a   Digital Transformation

"Clients do not come first. Employees come first. If you take care of your employees, they will take care of the clients."
— **Richard Branson**

Humans are experiencing technology's ability to transform life in ways which were unimaginable a decade ago. For example, online shopping has changed our behavior because we no longer need to physically go to the store. One can order almost anything from anywhere on their mobile phone. This serves as a reminder that digital transformation at an organizational level has an impact reaching far beyond technology and enterprise alone.

Customer-driven   digital   experiences   are   becoming prospective, leading plenty of marketers to bet everything on that progressing digital sector. Digital transformation and innovation demand high-powered capabilities on the technical side. However, what often goes unnoticed is that humans are still the principal factor in all of these changes. In practice, it is the effective digital skills of the employees that lead to digital transformation. Businesses must incorporate a new culture that uses digital technology advancement to empower people to learn, create new innovative solutions, drive change and bring disruption to the status quo.

Digital transformation can only occur when organization understands the impact of both internal and external individuals' have on the business. For instance, when a private citizen transacts with the local government, they are simply focused on their personal issue. By deploying a customer-centric digital transformation strategy, the government can drive a strategy that is not only outcome-oriented but also intuitive for users.

Platforms   can   handle   the   mechanics   of   managing information, but it is only through creating a culture of transformation amongst the individuals within the organization that the business can efficiently transform. This means breaking down departmental silos and encouraging staff to collaborate as they work towards the common goal of giving an exceptional experience to all stakeholders. Without the culture of change, a business might become digital but not have actually transformed. This will not be sustainable in the long run.

The human side of digital transformation needs to function efficiently, and organizing people is harder than organizing technology. A study by MIT and Capgemini which surveyed global 2000 companies, found that 90% of surveyed companies stated digital skills as a must for their future. However, 77% of those companies felt that those digital skills were missing in their enterprises. [74]

Companies should make digital training as a part of their operation. Up-skilling of employees should never stop for companies to evolve and innovate. The most fruitful programs will emphasize formulated learning opportunities, new content and assets, and the development of job experience.

Business technology leaders understand that digital economy requires a new workforce with new skillsets and new attitudes. However, organizations are struggling with skill shortages as the brightest talents are attracted to startups since millennials see a rapid change of workplace as the norm. Creating a liquid workforce means striking the right balance between the developments of a startup mentality, within the organization while simultaneously maintaining core business processes. This requires more than the correct engagement strategy for the digital natives, who soon will be a majority of the workforce. It requires major cultural re-engineering of organizations. The "born-digital" era, including human birth into the digital ecosystem, claims a world designed to accommodate its needs and desires around how work ought to be organized.

For a proper development of real-time communication in a distributed workforce, contractors must be provided with associated tools that are efficient across distances. An illustration at MindFlash allows us to understand better. The contractors were given, real-time access to each individual in their project team, through slack channels and Sococo spaces. This gave ability for immediate file and screen sharing. The implementations benefited both contractors and employees involved in the organization.

Companies need to know, how important feedback mechanisms and target success rates are, in handling liquid workforces. For instance, the Agile Scrum methodology, including fifteen-minute regular team meetings, periodic reporting of intermediate project outcomes and progress monitoring are all aspects of effective liquid team management.

Organizations often lack top talent with deep skills that are needed to successfully face the world's most challenging issues. In the sharing economy, governments and companies must work together to create a new social contract for those in today's fragile, flexible and innovative workforce. Leading companies are allowing anytime, anywhere working options using collaboration tools and cloud-based workflows which benefit both the worker and the business.

A more horizontal style of leadership in co-creating and managing is required to sustain a digitally skilled workforce and agile project models. With readily available data and artificial intelligence, more decisions will be made on the front lines where work transpires.

A dexterous workforce will only prosper in an organization that, confronts change, and equips to maneuver and adapt. It is that process which requires thorough oversight. Consequently, more organizations are investing in end-to-end workforce management solutions—such as those provided by Oracle, Workday and SAP to deliver key insights into workforce capabilities and readiness. This allows leaders and organizations to get more information about their workforce which they can use to optimize the organization's output.

Creating an agile workforce is challenging, but the rewards are boundless. By controlling the established workforce, businesses can be developed in ways that are smarter and faster than they ever envisioned. In the digital era, this constant growth and solid footing are more crucial than ever.

IBM is one of the companies developing a workforce that is capable of digitally changing the business landscape. IBM's strategy paper titled "Liquid" aims towards a more adaptable organization, one that is more suitable for the digital age. Only a small percentage of executives—who develop IBM's strategy and interact with clients—will retain the traditional steady jobs at the IT giant, while the rest are hired on a project-to-project basis for variable durations of time, depending on the workload. Workers can in turn offer their services on a platform analogous to the online auction house, eBay. Companies from all over the world can access these virtual stalls in search of freelancers.

Cloud computing and workforce networks are changing the way the contemporary organizations work. The traditional employment model—working nine to five and for years with the same employer—is on the decline everywhere. About 50% of the workforce in Germany work this way. The rest are specialists, casuals, or can be booked through temp agencies.

A new order is being established which is especially predominant in the IT sector. The relationship between the individual and the organization is the greatest challenge in the changed business world. Lynda Gratton, professor of Management Practice at London Business School has a beautiful analogy: *"It's been sort of a child-parent relationship, where the employee was the child and organization, the parent. Over time and with the new technologies, the relationship has shifted to an adult-to-adult connection."* She believes that people will be able to choose their own career and learning journeys.

The personal success of an individual depends on their professional digital fame. The people in talent cloud will post their career success and skills instead of their favorite music track or photos. The resulting CV will be a premise for future work applications for companies utilizing the liquid model.

Businesses are rapidly realizing that a liquid workforce is the new benchmark. Conventional methods that fall short are unable to adapt to changes in the digital world, and businesses are now learning that their workforce is another aspect of their business that can and should be a competitive advantage.

21

Winning Over The Global Crisis

"Everyone here has the sense that right now is one of those moments when we are influencing the future."
— **Steve Jobs**

Unprecedented times! Everyone is wondering how the world has changed in such a short period and how we are going to manage through this? How do we survive, recover, sustain and grow through this global crisis? What could you have done to have been better prepared?

The world has seen major global crisis right from the great depression of 1931, to the dot com bust in 2001, the financial meltdown in 2008 and the biggest of all, the Covid pandemic in 2020. These are times when leaders need to take care of the present, focus on recovery and prepare for resiliency in the future.

Over the last decade, technology has disrupted numerous businesses across all industries. We all have seen how the lockdown has decimated the traditional businesses and lives of each and every one of the seven billion global population. Factories have been shut down, offices and retail shops have been closed, hospitals are limiting themselves to urgent and emergency care. All non-essential products and services are unavailable. Unemployment is at its highest ever in history. Customer service contact centers have huge volume of calls, but very few agents to take the calls. All businesses are working from home. We all wish that our applications were all hosted on public cloud like Azure, AWS, Google cloud that come with many tools which help automate operations. This would have enabled a seamless "work from home." Many businesses have already made a decision that work from home will be the new normal. Most of your processes should be done through AI and digital workers rather than human workers. There are many intelligent solutions available driven by AI. Bringing in AI driven innovative solution is a must. So, before you automate "as is", stop and think. Ask yourself few questions.

1. Dream the customer experience you want to deliver, without technology constraints
2. Is there an "off the shelf" innovative AI based solution to enable the client experience? And you bet, there is!
3. Can I completely automate the solution? And you bet, you can! Don't listen to people who give you multiple reasons as to why it cannot be done

Businesses who have embraced innovative technologies have done extremely well. Numerous examples of winners should give us the motivation and sense of aggression for action. We all wish, we would have been more prepared earlier. The good news is that each and every one of us in our respective businesses can still leverage numerous innovations and the new technology, to get our businesses on track in a fairly short period. We need to prepare for the recovery and get the return on investment in less than six months. Here are few examples across industries that we can now at a near to zero upfront cost.

1. Call Center Automation

- **Problem**
  - Lack of seamless omni-channel and integrated solution
  - Legacy and outdated IVR and chatbots
  - Lack of attended automation
  - Use of legacy and cumbersome solutions hosted on proprietary data centers
- **Solution**
  - Omni-channel digital contact center
  - Automatic conversion of standard IVR system into conversation based smart IVR

- o Virtual Assistant capable of interacting through Phone, Web, Mobile, WhatsApp, FB Messenger, email, SMS etc.
- o Seamless transfer to live human agent with full context
- o Google Dialogflow, Text-to-Speech and Speech-to-Text
- o NLP, Interactive Forms
- o Multi-screen login integration
- o Cloud based WFM Digital Call Centers

- **Impact**
  - o Over 50% reduction in personnel cost and processing time reduced by over 5X (>80%)
  - o Go live in 4 weeks with 25% to 30% first call resolution and lower cost

- **Technology**
  - o Conversational AI based integrated omni-channel platform addressing customer journey, analytics, intent prediction and attended automation

2. Cloud Migration, DevOps, SecOps & WFH

- **Problem**
  - o Adopting and harnessing the power of the cloud is complex
  - o Current practice requires writing thousands of lines of code
  - o Opening a port for an application takes days
  - o Infrastructure Management requires two Desecops engineers for every 50 virtual machines
  - o WFH challenges

- **Solution**
  - o Cloud bots can manage the full Devsecops lifecycle starting from cloud infrastructure, application provisioning to CICD
  - o Includes all configurations in the cloud provider, Kubernetes and SIEM (security and incident management)
  - o Enable WFH and Change Management

- **Impact**
  - o 65% in operation savings
  - o 36% in application migration
  - o 15X increase in productivity
  - o 98% reduction in configuration errors
  - o Live in production in less than 2 weeks

- **Technology**
  - o DuploCloud - intelligent Devops bot, auto-generates the configuration which today humans are code manually. It

combines application specifications and compliance requirements with principles of a well architected framework to arrive at an end to end DevSecops implementation
- o These practices are used inside of Amazon, Microsoft and Google to scale each engineer to manage thousands of workloads

3. Unemployment Benefits Attended Automation
- **Problem**
  - o Setting up and managing employment benefits is a difficult, costly, time consuming, and error prone process
  - o Users are constantly unsatisfied with the level of service and timeliness of benefits
  - o Delivering benefits is an important part of government service to citizens
- **Solution**
  - o Multiple Screen Login and Integration
  - o Interactive Forms and Google Dialogflow
  - o Google Text-to-Speech and Speech-to-Text
  - o Natural Language Understanding
  - o No upfront cost – outcome based commercial model
  - o Work from home enabled cloud-based Call Center
- **Impact**
  - o 50% reduction in time to fill forms
  - o 90% reduction in claims processing
  - o Live in production in less than 3 weeks
- **Technology**
  - o Conversational AI based integrated omni-channel platform addressing customer journey, analytics, intent prediction and attended automation

4. Automated Digital Agent Alert Monitoring
- **Problem**
  - o Human agents required for monitoring, analysis, follow-up with users, escalation to administrator and policy compliance
  - o Reporting devices that are not connected
  - o Reporting of rouge applications
  - o Reporting of cyber-score
  - o Detecting fails and auto-remediation
- **Solution**

- o Architected compliance-as-a-service
- o Digital agent alerts and IOT data analytics monitoring & policy compliance
- o Auto-remediation driven by AI and intelligent automation, based on defined policies
- **Impact**
  - o 100% intelligent automation with self-adjudication based on defined cybersecurity policy
  - o 90% cost reduction
  - o Live in production in less than 2 weeks
- **Technology**
  - o Compliance-as-a-Service
  - o User end point management of devices
  - o End to end intelligent automation and self-adjudication

5. Mortgage Loan Risk Management

- **Problem**
  - o Mortgage loan performance is difficult to forecast
  - o Typical models have at best a 65% accuracy
  - o Inaccurate forecasting can lead to significant losses and opportunity cost
- **Solution**
  - o Proprietary data science, ML and AI based technology
  - o AI/ML based on 1.6 million mortgages from 23 years
  - o Simulation of multiple servicing actions for high risk loans
  - o Generate new performance, cash flow forecasts and net present value for each servicing action
  - o Provide a list of actions that yield a statistical value increase, in order of value gain
- **Impact**
  - o 96.5% accuracy in loan forecasting
  - o Go live in less than 2 weeks for technology implementation and less than 4 weeks for servicing
- **Technology**
  - o Data science analytics based on AI/ML from 1.6 million mortgage loan data from 23 years of transactions
  - o Intelligent Automation of mortgage servicing actions

6. Healthcare Claims Automation

- **Problem**

- o   Workers Compensation is froth with issues and fraud
- o   Fee for service leads to overprovision, inefficiency, uncontrollable health expenditure, serving the physician interest
- **Solution**
  - o   Patient outcomes and value-based approach
  - o   Evidence based medicine
  - o   Improve preventative care
  - o   Reduce hospital re-admission
- **Impact**
  - o   Evidence based treatment
  - o   50% lower cost to payor
  - o   Significant increase in patient satisfaction
  - o   Go live in less than 4 weeks
- **Technology**
  - o   Patent pending proprietary software platform
  - o   Data science, AI/ML and evidence based medical data
  - o   Provider intelligence and Intelligent referral
  - o   Intelligent Care Trend

7. Automated Check Processing

- **Problem**
  - o   Manual entry and verification of name, date, amount and signature from check image
  - o   Time consuming processing, by costly bank employees
  - o   High volume repetitive process and prone to errors
- **Solution**
  - o   Check data (date, amount, beneficiary name and signature) is extracted by a proprietary AI based OCR/ICR technology
  - o   Intelligent Automation for check processing, payment, reconciliation, validation of signature / check number vis-a-vis A/C number from CBS system
  - o   Validation of all banking rules for check clearance
- **Impact**
  - o   Over 50% reduction in personnel cost and processing time reduced by over 5X (>80%)
  - o   Go live in less than 4 weeks
- **Technology**
  - o   Proprietary Intelligent OCR/ICR
  - o   Proprietary computer vision & deep learning

8. Supply Chain Automation

- **Problem**
  - o Inefficiency, wastage and spoilage in operations
  - o Lack of audit trail for compliance
  - o Lack of visibility and intelligence across supply chain
  - o Lack of integrated and continuous monitoring of location, ambient conditions (temperature, humidity, pressure, shock, light) both indoors (manufacturing facility, warehouse) as well as in transit through various modes of transport
  - o Alerting stakeholders on business-specific exceptions
- **Solution**
  - o Long-life sensor tags, gateways, and cloud-based AI/ML powered control tower
  - o Real-time information from objects and processes using hard and soft sensors
  - o Streaming data analysis, application of business rules and automated notifications
  - o Digital twin platform
- **Impact**
  - o Lower supply chain management cost by over 90%
  - o Go live in less than 2 weeks
- **Technology**
  - o IOT based AI/ML real time continuous proprietary technology providing an integrated view of all data, alerts and automated exceptional handling

9. Supply Chain Digital Agents

- **Problem**
  - o Manual login to customer portals to update shipping and delivery information
  - o Data capture from ERP for each customer
  - o Over 100 customer portals to be updated daily
  - o Manual entries resulted in errors and discrepancies
  - o No real-time aggregation of future demand
- **Solution**
  - o 100% automated update of data from ERP to customer portal
  - o Collection of real time aggregated demand data from customer portal and distribute to respective departments for production planning
- **Impact**

- o 80% TAT reduction
- o 100% data accuracy
- o 100% automated
- o Live in production in 3 weeks
- **Technology**
  - o Process redesign and intelligent automation

10. Automated Online Product Listing

- **Problem**
  - o Numerous data points and steps for product listing on Amazon
  - o Manual and time-consuming entry of product details
  - o Numerous templates, multiple product categories and series of sub-categories
  - o Multiple seller account creation, product uploading while keeping acceptable formats and information
- **Solution**
  - o The intelligent bot registers new seller, logs in, captures product information from the various source files
  - o Validation of all data points and rules for product listing
  - o Automated entry of data in Amazon template, automated upload in portal and go live of product listing
- **Impact**
  - o Over 50% reduction in personnel cost and processing time reduced by over 5X (>80%)
  - o Go live in less than 4 weeks
- **Technology**
  - o Intelligent OCR/IMR and intelligent automation

11. Banking - KYC/AML

- **Problem**
  - o Manually reading the data from input files from different sources to know customer information
  - o Validating every field in customer profile through research
  - o Significantly higher number of AML false positives and SARs
- **Solution**
  - o Automated data collection, research and validation
  - o AI/ML based insights and transaction analysis and reporting of AML suspicious activity (SARs)
- **Impact**

- o 70% reduction in average KYC time and cost
- o 80% reduction in AML/SAR false positives
- o Live in production in less than 3 weeks
- **Technology**
  - o AI/ML and insights based proprietary technology

12. SAP Migration and Operation

- **Problem**
  - o All clients must migrate to S/4 HANA by 2027
  - o Only 7% clients migrated so far; critical lack of project resources/talent
  - o The market needs 140 migrations a week to hit SAP's mandate
  - o SAP migration and operation costs from millions to billions of dollars/year
  - o Migration is difficult, costly and risky to core business functions
- **Solution**
  - o 35% lower SAP migration & operation cost, risk and effort
  - o 130+ SAP specific automations – sellable per unit or packs
  - o Mapped to SAP ACTIVATE implementation methodology
- **Impact**
  - o Over 35% cost reduction
  - o >50% reduction in time to implement
  - o Go live for start of migration in less than 4 weeks
- **Technology**
  - o Process re-engineering, pre-build migration assets and intelligent automation

These are just few examples of AI driven intelligent automation solutions that businesses are deploying now. Tholons (www.tholons.com) and world's leading consulting and technology companies are engaged with visionary businesses to deploy these solutions. You, as a leader have the fiduciary responsibility to your employees, customers, families and your business to act NOW!

22

Digital Locations

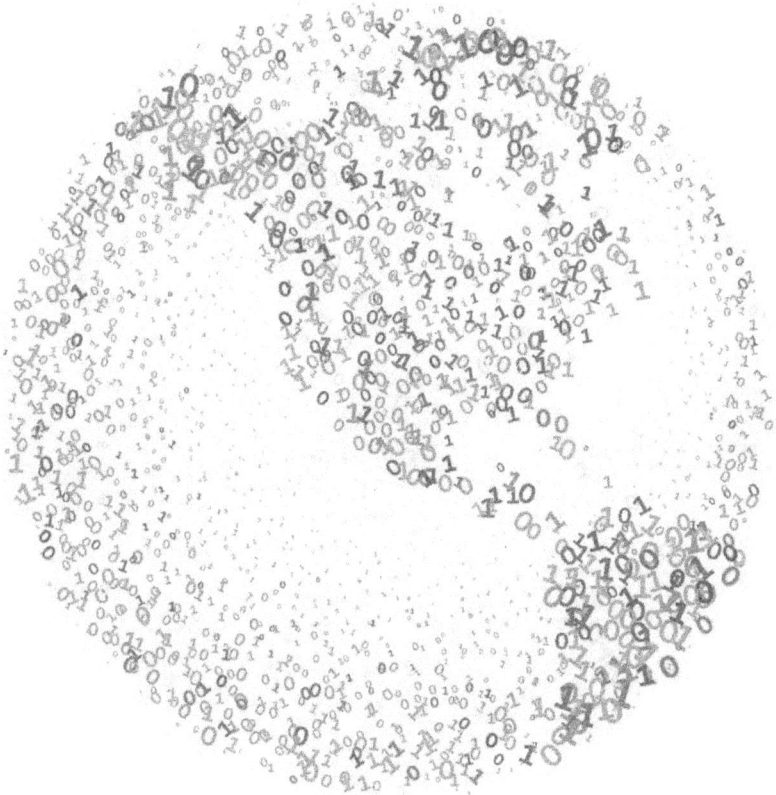

Argentina

"Every once in a while, a new technology, an old problem, and a big idea turn into an innovation."
— Dean Kamen

Argentina officially the Argentine Republic is a country located in the southern half of South America. The country has a highly educated workforce and is a global exporter of agricultural products, energy, and mineral resources. The nation's considerable internal market size and a growing share of the high-tech sector has made it one of the world's top developing nations.

Argentina has a mixed economic system which includes a variety of private freedom, combined with centralized economic planning and government regulation. Argentina ranked in 30th position according to the International Monetary Fund Nominal GDP Ranking 2019 report. Benefiting from rich natural resources, a highly literate population, a diversified industrial base, and an export-oriented agricultural sector, the Argentinean economy is emerging.

The Startup Ecosystem 2019 rankings placed Argentina at 44th position globally. The country has more than a thousand startups, and Buenos Aires is its major startup hub. Argentina has the potential to be the innovation hub of Latin America, which is evident from the nation's history of producing the highest number of successful companies in the continent. The country has hosted four Unicorns, namely Mercado Libre–e-commerce platform; OLX–online classified ads company; DESPEGAR–IT and software development; GLOBANT–online travel agency. Successful startups like Investo Land, Relojes Especiales, Smart Labs, Live Nation, iAgriLink give Argentina an edge over its regional competitors.

According to Tholons Services Globalization Index 2019 rankings, Argentina is positioned 10th among top 50 'Digital Nations'. The capital city Buenos Aries featured in the report at 10th position in the 'Top 100 Super Cities'.

The presence of a large number of accelerators and incubators has played a major role in keeping up the pace of startup growth in Argentina. It is home to more than 150 startup accelerators and incubators, 96 co-working spaces, and 28 Venture Capital funds. NXTP Labs and Wayra. NXTP Labs are the leading accelerators in the city, that have made over 183 investments and are raising a large venture fund aimed at providing access to scale. IncuBAte–global incubator fund, provides up to USD 30,000 equity-free seed funding for early-stage startups, and assist young talents in registering, and launching businesses locally.

The Argentinian startup ecosystem has grown at an unprecedented rate. Academia Buenos Aires Emprende and StartUP BA are some programs offering entrepreneurs access to training and mentorship. Programa Emprende INNdustria focuses on providing support to industrial startups. Consejo Consultivo Emprendedor and Mendoza Emprende are some of the private and public programs organized to boost entrepreneurship culture. LAVCA's (Association for Private Capital Investment in Latin America) 2018 pooled in international investors with USD110M across 27 Argentine startups.

The Argentinian government is investing in venture funds and aligning policies to make it easier for people to start a new enterprise or invest in companies. Several programs supporting entrepreneurial activity are flourishing in the country's biggest startup hubs, Buenos Aires, Cordoba, and Mendoza. The government's Entrepreneurship Law seeks to open up Argentina's markets and attract investment from abroad and reduce challenges in initiating a new business. It also contributes to 13 accelerator programs, enabling public crowdfunding, and the creation of a government-run fund. Argentina Innovadora 2020: National Plan of Science, Technology, and Innovation is the instrument by which the Ministry sets guidelines for science, technology and innovation policy in the country. It aims to continue with the growth and consolidation of areas considering strategic pillars of national development. The plan focuses on strengthening innovation by training high-quality human resources and promoting the development of entrepreneurial culture and innovation.

Argentine entrepreneurs are building a vibrant entrepreneurial ecosystem. The potential to foster successful companies is evident from the number of startups and the diversity of unicorns the country has produced. Even though Argentina has faced a severe economic crisis for almost two decades, the startup ecosystem didn't stagnate and continued to progress. The recent reforms in policies and regulations reflect the commitment of the government to establish Argentina as an innovative and high-caliber tech startup hub in Latin America.

Australia

"Computers are incredibly fast, accurate and stupid; humans are incredibly slow, inaccurate and brilliant; together they are powerful beyond imagination."
— **Albert Einstein**

Australia, officially the Commonwealth of Australia, is the smallest continent that lies amidst the Pacific and Indian ocean. It is a sovereign nation comprising the mainland of the Australian continent and one of the largest countries in the world. The country has always relied on its natural resources; therefore, a more expanded economy is critical to secure Australia's steady growth. The entrepreneurial soul, inquisitive personality and logical ability of Australia allow it to stand out in a good position globally.

Australia is a highly developed country. It is a mixed market economy which includes private freedom, combined with centralized economic planning and government regulation. According to the World Banks Nominal GDP Ranking Report 2019, the country ranks in 14th place. Australia is highly urbanized and has stamped 28 years of uninterrupted annual economic growth. Propelled assembling, innovation, logical advancement and creating future talent remains a focal part of the Australian innovation industry.

According to AngelList, Australia is home to more than 9,000 startups, and ranks 5th position globally among 202 countries, based on the startup ecosystem. Information Media and Telecommunications, Finance and Insurance, Professional, Scientific and Technical Services are sectors in which the majority of Australian startups are innovating. Sydney, Melbourne, and Brisbane are the cities with the most vibrant startup ecosystems in Australia. According to Tholons Services Globalization Index 2019, Australia ranks 16th position among the top 50 'Digital Nations'. Sydney (24) and Perth (59) are among the top 100 'Super Cities' in services globalization.

Australia is setting up to succeed in this highly innovative and competitive global economy. There are 83 accelerators and incubators, 233 venture capital funds in Australia. Innovacorp, Muru-D, CyRise, Startmate, The Actuator are some of the major accelerators. The startup ecosystem is supported by a considerable number of venture capitalists and angel funds. Dominant Digital, Optus-Innov8 Seed, ANZ Innovyz START, Sydney Seed Fund, Black Citrus, Tank Stream Ventures, Square Peg are some of the early-stage investors. BCG Digital Ventures, Antler, Blackbird Ventures, Yuuwa Capital, Starfish Ventures, Southern Cross Venture Partners are the major venture capital firms in Australia.

The government's innovation statement laid a solid foundation for startup-led innovation. StartupAUS is a government-funded organization launched in 2013, focused to build Australia as one of the best places to establish and develop a tech startup globally. They are improving the regulatory environment building a case for the right sort of government support for a fast-growing tech sector. Many events are organized by public and private in collaboration to promote the Australian startup scene. Lean Startup Melbourne is the biggest and most popular startup event conducted every month in Australia. Startup events like Startup Grind, Silicon Beach Drinks, Startup Weekend and Startup Healthtech, are organized every month that focuses on workshops to educate and inspire the entrepreneurial community.

The Australian government is propelling the growth of startups by providing several incentives in the form of grants, tax benefits, and reimbursements. Governments Entrepreneurs

Programme offers commercialization fund and business growth grants to budding entrepreneurs. If the annual turnover of a company is less than USD 20 million, they can claim a 43.5% refundable tax offset against R&D expenditure that is up to USD 100 million. Austrade Landing Pad is a website that provides market-ready startups with potential for rapid growth, a cost-effective option to land and expands into one of five world-class innovation hubs–Singapore, Berlin, Shanghai, Tel Aviv and San Francisco.

There is no question that the competitiveness of the global economy now is tougher than ever. For a country like Australia, businesses have to evolve and innovate to remain productive and flourish. Innovation shouldn't be something exclusive to startups and tech companies, it should be applicable across the economy. However, with the latest developments in the Australian entrepreneurial structure and government initiatives, the future of the Australian startup ecosystem looks bright.

Brazil

"Computer technology is so built into our lives that it's a part of the surround of every artist."
— **Steven Levy**

Brazil, officially the Federative Republic of Brazil is the largest country in South America. With over 208 million people, it is the fifth-largest country by area and the fifth most populous in the world. It is one of the world giants of mining, agriculture, and manufacturing, and has a strong and rapidly growing service sector. According to the International Monetary Fund outlook 2019 report, the nation ranks ninth largest in the world by nominal GDP and eighth by purchasing power parity (PPP) measures.

Brazil is home to more than 10,000 startups and ranks 37 globally among 202 countries, based on its startup ecosystem strength. The cities with the most vibrant startup ecosystems in Brazil are Sao Paulo, Rio de Janeiro, and Belo Horizonte. Startup Ecosystem Ranking states Sao Paulo is leading in the number of new techs startups followed by Rio de Janeiro with a total of six thousand established tech companies across the country.

Brazil is at the forefront of startups with 7 unicorns, and 6 among them emerged in the year 2018. They are 99–transportation network company; PagSeguro–e-commerce service; Nubank–financial services; Arco Education–edutech startup; Stone–

payments company; Brex–fintech startup; and Loggi–food tech. As of 2019, Brazil has about 369 business incubators, 90 tech park initiatives, 238 co-working spaces, and 35 accelerators involving 6,500 innovative firms.

According to Tholons Service Globalization Index 2019 report, Brazil ranked second among the Top 50 Digital Nations and Sao Paulo at third among the Top 100 Super Cities. Curitiba (33), Rio De Janerio (42), and capital city Brasilia (68) are also in Tholons Top 100 Super Cities list.

Brazilian startups have attracted foreign investments and are creating new jobs for skilled locals. Agricultural technology startups in Brazil are getting investment from Monsanto–a leading biotech company partnered with Brazil Microsoft, aiming to bring innovation to the agricultural sector. The selected startups receive initial funding up to 1.5 million BRL (USD 459,000) for early development. Fintech and B2B services are also in the lead. Finnovista–an impact organization that accelerates the development of technological companies, states that Brazil is in the leading position as the largest fintech ecosystem in Latin America with over 370 startups. The report found payments, lending, personal finance, and insurance are three promising startup sectors. Venture capital investments are also on the rise, with a 35% annual growth rate from which 41% were directed to early-stage enterprises. Brazil is estimated to generate potential revenue of about USD 24 billion by 2026 and intends to chase a position in the high-tech industry with its startup companies.

Acelera MGTI, a local state-run startup accelerator program, reports the Brazilian tech market contributes 5% to the country's GDP, i.e., over USD 100 billion. With 46% share of the market, Brazil's tech sector ranks seventh in the world and first in investments, in Latin America. ABRAII – Brazilian Association of Companies Accelerator for Innovation and Investment, states the country has more than 40 programs on full throttle to foster startup culture. A lab is setup in Silicon Valley by Brazil's largest bank–Banco do Brazil, to identify and discover new fintech solutions that can transform the banking sector. The Plug and Play Tech Center is a business accelerator in Silicon Valley that is creating an ecosystem for startups and hosts the Banco do Brazil Advanced Laboratory.

Hackathon events and innovation programs are also conducted for employees and students from local tech institutes.

There are 122 incubators from public universities and the majority of them work in technology-focused sectors. Google opened its R&D centre in Belo Horizonte in 2005 to leverage the local vibrant innovation environment. Two years later, Stanford University started the first edition of its Entrepreneurship and Innovation program in Belo Horizonte.

To promote startup culture in Brazil and attract foreign investments, a total or partial exemption from duty, excise tax and social contributions on imported equipment is granted to investment projects. The Informatics law allowed market entry for global organizations into Brazil. The law permitted global organizations to sell their products and services in Brazil. It also incentivized them to invest in Research and Development. Investments in the research sector are encouraged by the law which has led to the making of many technological parks in the nation. Income tax exemptions or reductions are also available for companies set up in specific regions within Brazil.

Brazil is not dubbed the Powerhouse of Latin America without a reason. Its massive population and untapped potential across various industries offer the allure of great reward to entrepreneurs and investors. Many startups are carving niches for themselves in the domestic market and have gone on to been backed by prominent venture capitalists. At the same time, traditional investors, companies, and governments are showing interest in ecosystem participation through mentorship, funding and partnership opportunities. With its vibrant startup ecosystem, interest in innovative ideas, and investments Brazil can be regarded as one of the global leaders in startup innovation.

Canada

"There is far more opportunity than there is ability."
— **Thomas Alva Edison**

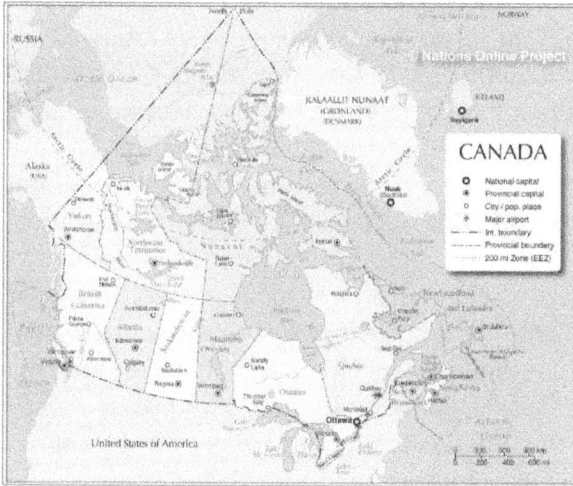

Canada is located in the northern part of North America and the world's second-largest country by total area. Technologically advanced and industrialized, Canada is heavily reliant upon its abundant natural resources and trade. It is one of the world's top ten trading nations with a highly globalized economy.

Canada is a mixed economy and is ranked the world's tenth-largest economy as of 2019. With a nominal GDP of approximately USD 1.73 trillion. Its competitive cost, ability to attract talent and proximity to the largest economy (US) gives it a significant advantage. Canada ranks 22nd position in World Bank Group Ease of Doing Business report 2019. It is positioned well to challenge and step up to disrupt established tech leaders.

Canada is determined to establish itself as one of the top startup hubs globally. Canada has ranked 5th place in Startup Blink Ranking of Countries for Startup Environment 2019 report. Vancouver and Toronto are the two major startup hubs of the nation. Vancouver is referred to as the 'Silicon Valley of the North' with over 1,500 startups and three unicorns in the city itself. The city has the most thriving startup culture, infrastructure, and quality of graduates matched by no other city and is in "the race" to become a major business hub. Toronto, the financial capital of Canada, is also making inroads for innovation. Its location as a base for many international financial companies gives it an advantage in both connectivity and access to funds.

In Tholons Services Globalization Index (TSGI) 2019, Canada ranks at 4th position among the Top 50 Digital Nations. Canada has five of its cities in Top 100 Super Cities–Toronto at 6th, Calgary at 36th, Halifax at 49th, Vancouver at 60th followed by Montreal at 63rd position in the list. In the same report, Canada also features in the Top 25 Potential Digital Leaders with Toronto ranked 13th and Montreal bringing up the rear in the 21st position.

According to AngelList Canada's fast-growing startup ecosystem has 6 unicorns and over 16,000 startups. OneConnect–financial technology services company; Shopify–e-commerce company; Kabam–gaming platform; Kik Interactive–social networking app; Slack–cloud-based collaboration software; Hootsuite–social media management platform; Avigilon–surveillance solutions are the unicorns of Canada. Canada has over 330 co-working spaces, 150 accelerators and incubators, and is a natural hive of entrepreneurial activity.

Canada is committed to foster its startup ecosystem. Universities are also providing infrastructure and various other supports for the growing entrepreneurship sector. Digital Media Zone is a combined incubator/accelerator program by Ryerson University that has assisted more than 130 companies. It has organized a working committee on innovation, comprising VCs, startup CEOs, tech executives, and mentors. This committee formulates policies and connects business leaders and government to ensure a very supportive ecosystem for business.

The Canadian government has started many initiatives to bolster a fledgling tech ecosystem. It has developed tax incentives, loans, grants, wage subsidies, and even offered visas to qualifying immigrants who want to plant their startup on Canadian soil. MaRS Discovery District is one such initiative launched back in 2005 supporting Canada's most promising startups—helping them grow and create jobs. It provides essential skills to entrepreneurs through its venture program. Industrial Research Assistance Program is a government-backed program offering budding entrepreneurs the mentorship and funding they need to get their startup off the ground. IBI (Investing in Business Initiatives) and Smart start are among various initiatives started by the government with an interest to promote investment in early-stage companies.

The availability of venture capital in Canada has changed dramatically over recent years. Market reach is Canada's strongest factor and the city is experiencing a good inflow of foreign investments is rising encouraging venture investments in startups even further. The Ontario Municipal Employees Retirement System (OMERS) deployed USD 180 million in early-stage startups between 2011 and 2014, while BDC Capital launched its IT Venture Fund II, a fund worth USD 150 million.[75] Generous tax credits for new businesses and the expansiveness of the city make the cost of infrastructure and essential capital low. British Columbia, Quebec, and Ontario are progressive provinces leading the charge in venture investments.

Every successful startup needs a boost, whether that comes from mentoring, venture capital, an incubator, a grant, loan, cultivation event, or all the above. As Canada and its provinces continue to promote and carefully expand their startup culture, there are plenty of resources available for the savvy entrepreneur who knows where to look. Investments and rising industries within the world of technology are booming in Canada. With a system that rewards innovation, encourages collaboration, and supports diverse, global entrepreneurship, it's no surprise that Canada has a bright startup future ahead. As long as entrepreneurs and government programs continue to thrive, the country will remain a beacon of hope for entrepreneurs.

Caribbean

"We know that, when it comes to technology and the economy, if you're not constantly moving forward, then-without a doubt-you're moving backwards."
— **Bill Owens**

The Caribbean known is a region of the Americas situated largely on the Caribbean Plate and has more than 700 islands. It has an educated multilingual workforce with sophisticated financial systems and is a short hop to the United States, Mexico, and other large markets. The region has great economic potential growth and opportunities to further develop its services, logistics, agriculture, creative and digital sectors.

Jamaica

Jamaica is an island country situated in the Caribbean Sea with a population of 2.8 million. It is the third-largest island of the Caribbean and the third-largest English-speaking country in the region. Tourism employs 25% of the Jamaican workforce and represents 5% of the country's GDP.

Jamaica has been ranked 50th among 'Digital Nations' in Tholons Services Globalization Index 2019. Kingston, the capital city is also ranked 99 among the Top 100 'Super Cities'.

StartupBlink ranks Jamaica 88 globally among 202 countries, based on its startup ecosystem. The cities with the most vibrant startup ecosystems are Kingston and Montego Bay. There are over 100 startups in Jamaica as per AngelList. The following are the top startups of the country: JamGora–online marketplace; MediRevu–health application; Radial–music platform; Ja++–digital marketing solutions; and RevoFarm–business analytics platform.

   The Jamaican government is implementing several policy reforms to promote the growth of entrepreneurship and innovation. JAMPRO, an agency of the Government of Jamaica's Ministry of Economic Growth and Job Creation, promotes business opportunities in export and investment to the local and international private sector. Jamaica is taking strides to become an international hub for digital innovation and the Silicon Valley of the Caribbean. StartUp Jamaica–an initiative supported by the World Bank is designed to help young, savvy, digital entrepreneurs find investors such as Oasis 500 and tap into the global demand for mobile applications. It's a key element to turn the potential of Jamaica's youth into a real employment opportunity for them and future generations. The Jamaican startup ecosystem is growing fast and is truly a GREAT place to build startups. Many aspiring entrepreneurs are utilizing the local resources available to foster the nations startup, innovation, and entrepreneurship space.

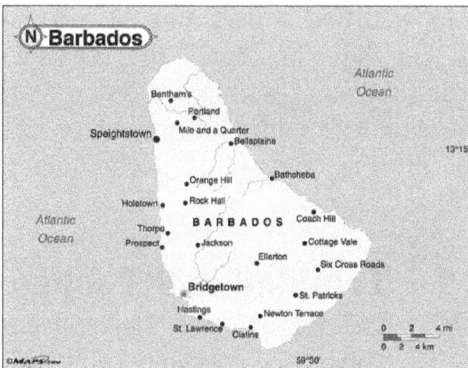

Barbados

Barbados is an eastern Caribbean island and an independent British Commonwealth in the Atlantic Ocean. It is an open market economy and competes globally in several industry domains. Barbados is ranked 53rd place in the world in terms of GDP (Gross Domestic Product) per capita. The country's top-level educational system has yielded a highly intelligent workforce with an abundance of professional as well as skilled labor. It has a highly educated workforce with a literacy rate of 99.7%. Barbados has up to 40 startups according to AngelList. Capital city Bridge town and Welches are the 2 major innovation hubs of the island. The most popular startups in Barbados are MediRevu–MedTech app; Skopeout–web application; and EGO Scholar Media Services–media services provider.

Many programs are launched in the country to promote the startup ecosystem. The Barbados Entrepreneurship Foundation (BEF) is a catalyst for bringing up positive change in the economy through the promotion and development of entrepreneurship in the island. There are multiple entities in Barbados which seek to meet the needs of entrepreneurs and the foundation works effectively to manage the island's limited resources.

To increase exports, employment, and help competitive businesses through new investments, the Barbados government has created an incubator program called BIDC. The primary objective of BIDC is to improve the chance of success, mainly intended for startups and upcoming businesses by ensuring an incubation period of three years. Under the incubator program, the Barbados Small Business Centre for clients provides workspace and workshops and in-house consulting services by BIDC Business Development Officers rendering professional support to startups.

The tech startup companies in Barbados are experiencing major difficulties in scaling their local businesses because of the small market, limited access to qualified coders and the chronic absence of venture financing. On the other hand, the Barbados government is actively supporting the local startup ecosystem both politically and financially. The country's educated population and affordable infrastructure are driving forces of technological innovation.

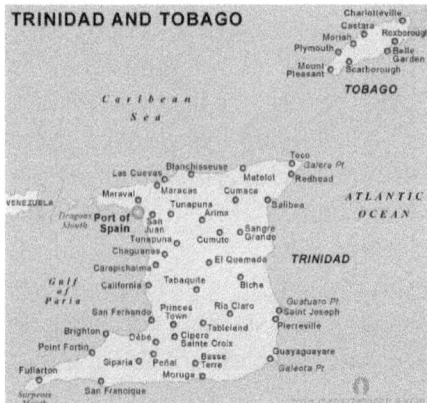

Trinidad and Tobago

The Republic of Trinidad and Tobago is an island country in the Caribbean located at the southernmost area consisting of the main islands Trinidad and Tobago, and numerous smaller islands. It has the third-highest GDP per capita based on purchasing power parity (PPP) in the Americas after the United States and Canada. The country is recognized by the World Bank as a high-income economy and is the most developed nation in the Caribbean.

According to Tholons Services Globalization Index 2019, Trinidad and Tobago have been ranked 49th among 'Digital Nations'. AngelList lists over 50 startup companies in the country. TapTag—web application and Way share—travel planner are the top startups of the region.

The government has started taking important initiatives to promote startups and SME's in the nation. Its goals are centered towards providing a unique combination of infrastructure, business development support, operational and financial assistance that will reinforce the growth and success of upcoming and existing micro and small enterprises (MSEs). The Ministry of Labor and Small and Micro Enterprise Development (MLSMED) created a project called the National Integrated Business Incubation System (IBIS). It provides seed capital for clients ranging up to USD50,000. uSTART is a university accelerator and incubator of Trinidad and Tobago functioning as an engine of innovation and enterprise. It supports the entrepreneurial mindsets of both students and staff of the university fostering the islands startup ecosystem.

Many events promoting startups are being organized in the Port of Spain, the capital city. Startup Weekend to Improve Lives—Trinidad and Tobago co-sponsored by the Inter-American Development Bank (IDB), is one among such events. The event provides startups an opportunity to pitch their ideas, form a team, build a product.

Though the Caribbean startup scene is in its developing stage, companies have always performed an excellent job in terms of innovation, creativity and inspiring young entrepreneurs. The government is making efforts to promote the innovative startup ecosystem. The Caribbean region has the potential of becoming the best choice for global investors to look for budding entrepreneurs and technology startups.

Chile

"I have not failed. I've just found 10,000 ways that won't work."
— **Thomas Edison**

Chile, officially the Republic of Chile is an American country, located in the continent of South America. With a populace of just 18 million, it is an unlikely entrepreneurial center. The high-income economy and living standards have made Chile most economically prosperous and socially stable nations in the continent.

Chile has turned out to be one of the swiftest developing economies in Latin America. The most striking characteristics of Chile is its political and financial growth. Chile has been ranked 56th in ease of doing business by World Bank in 2019. It is famous for its high level of innovation and a healthy economy. The Global Competitiveness Report for 2019 ranked Chile as being the 33rd most competitive country in the world and the first in Latin America. However, it is a genuinely modest place to live and work.

The startup ecosystem in Chile is rightfully called the "Chilecon Valley." According to AngelList, the country has over 1,500 startups. Santiago—the capital city, is one of the leading startup hubs globally and has established 1000 startups. Networking ties of firms like Chile Global Angels, Austral Capital and Magma Partners which are based out of the city and are naturally strong nearby. The Startup Ecosystem Rankings report 2019 ranked Santiago, first in Chile and 59th globally.

According to a report by Gust and Fundacity, total investment for startups in Latin America stood at USD31,563,841 and Chile topped the list with USD 15 million funding. Chile is among the global Digital Innovators and is ranked 12th in Digital Nations list according to the Tholons Services Globalization Index– 2019. Santiago featured in the Top 100 Super Cities report and is placed at 15th position.

Chile is quickly gaining a reputation for becoming Latin America's startup hub, with businesses creating global connections and providing valuable employment resulting in an investment boost across the country. The popular startups in Chile are ChileTrabajobs – job portal platform; Fintual – fintech solution; Cornershop – online marketplace; Portal Educativo – Edutech platform; GoPlaceIt – real estate platform; SaferTaxi – transportation network, and has produced a sole unicorn Crystal Lagoons–a design and construction firm.   Chile consists   of   over   20   accelerators and incubators supporting around 450 startups per year, more than 10 angel investors and family offices providing up to half a million dollars in investment.

The Chilean government initiated more than 50 programs to encourage new businesses. As indicated by the Brookings Institute, nearly 200,000 Chileans have gotten an advantage from government-supported business enterprise programs. Startup-Chile is one of the most successful initiatives launched in 2010 to change the nation's culture towards entrepreneurship. The program helps entrepreneurs to scale up their business ideas by awarding the selected few an equity-free government grant of 20 million pesos. Foreign entrepreneurs are expected to contribute 10% of the total grant's value and are incentivized to employ local talent. More than 1,300 small businesses are supported by seed accelerator and about 50 nations have replicated this publicly funded program. Startup-Chile is also offering an additional funding of USD10,000 to companies willing to set up startups outside of Santiago.

The rise in entrepreneurial endeavors in the country, universities are adding rigorous courses to their Business and Entrepreneurship program. Chrysalis, a state-upheld business incubator associated with the Pontifical Catholic University of Valparaíso was launched in 2009. The organization leads regional efforts to make entrepreneurship and innovation the engine through which the country can develop its potential to achieve growth. CORFO which is the principal public agency in charge of promoting the innovation in all types of enterprises, both consolidated and new enterprises, also has important lines of support intended for research centers.

FDI (Foreign Direct Investment) inflow values are particularly increasing in the nation. This is seen as a sign of increasing investor confidence, and service providers; in particular, looking towards Chile for globalization services. Foreign investors are encouraged by laws and regulations with fewer restrictions upon FDI in Chile., as well as the exemption from value-added tax. It is in these forms of institutional support that Chile maintains its distinct advantage over alternative locations. This is further enhanced by strong institutional support from the government through incentives offered for foreign investors.

Chile is said to be not only Silicon Valley of the region, but likely continue to grow and morph into Latin America's Singapore; a place of success, innovation, ingenuity, and profitability. Chile has a solid, sound and stable economy where the government can address business enterprise activities in a protected and innovative way. Moreover, organizations in Chile still rely on their natural resources. With more technological advancement and innovation there is a wide range of opportunities to foster startup growth in the nation.

Colombia

"The number one benefit of information technology is that it empowers people to do what they want to do. It lets people be creative. It lets people be productive. It lets people learn things they didn't think they could learn before, and so in a sense it is all about potential."
— Steve Ballmer

Colombia, officially the Republic of Colombia is situated in the north of South America. According to the 2019 World Bank Doing Business Report, Colombia ranked 65 out of 190 countries with a score of 69.24. It has a literacy rate of 94.65% and the most skilled and competitive labor force worldwide. The Global Competitiveness Report 2018 ranked Colombia's financial market development as the strongest pillar of its economy, ranking it 60 out of 140 countries.

AngelList lists up to 2000 startups in Colombia. According to Startup Blink Ecosystem Ranking 2019 report, Colombia ranks 34 globally among 202 countries. The most popular startup industries in the country are mobile, e-commerce and social media. Bogota, Medellin, and Barranquilla are closely competing to be the startup hub of the country. The most successful Colombian startups are Eforcers S.A.–cloud computing innovation company; Credivalores–consumer loan provider; Platzi–app development; and Liftit–an online platform for on-demand short-haul trucking. The country has so far produced two unicorns: LifeMiles–frequent flyer program;

Rappi–on-demand delivery service. The country has 53 accelerators and incubators, 31 venture capitals, and 13 co-working spaces.

Tholons Services Globalization Index 2019 ranks Colombia as 11th in Top 50 'Digital Nations', and Bogota (37th), Medellin (53rd), and Cali (100th) are also among the Top 100 'Super Cities'.

The ever-increasing incentives for new businesses and investors, such as free trade zones and tax deductions have accelerated the investment in technology projects and have made the country even more appealing. The free trade zones regime grants 100% of investment deduction to companies thriving to improve the startup environment and preferential 15% tax for companies rooted to establish a new business.

The government has initiated many programs and events to promote the growth of new ventures. The Apps.co initiative focusing on tech-based startups is a program designed by the Ministry of Information Technology and Communications (MinTIC), to promote and enhance the creation and consolidation of businesses and initiative. It has awarded USD 33 million in funding to accelerators and university partnership programs. INNpulsa, is a business growth management unit of the Colombian government, launched in February 2012 with a vision to make Colombia one of the top three most innovative economies worldwide by 2025. It promotes entrepreneurship, innovation, and business development awarding grants of up to USD 800,000 to investor groups establishing services in Colombia.

Colombia may not be at the forefront of the growth of dynamic and rising startup culture, but taking account of the current developments with regard to innovation makes sense. The country is rapidly becoming the new home of innovation, competitiveness and international growth. While Colombia is still a reasonable way off from garnering the success and attention of the aforementioned hotbeds, an ardent group of entrepreneurs is prospering inputting the necessary work to create an atmosphere conducive to successful startups in the region.

Costa Rica

"The most beautiful thing we can experience is the mysterious. It is the source of all true art and science."
— Albert Einstein

Costa Rica, officially the Republic of Costa Rica is a sovereign state in Central America, with a population of 4.8 million and well-known for its highly educated workforce. The country has a mixed economic system, which includes private freedom, combined with government regulation and centralized economic planning.

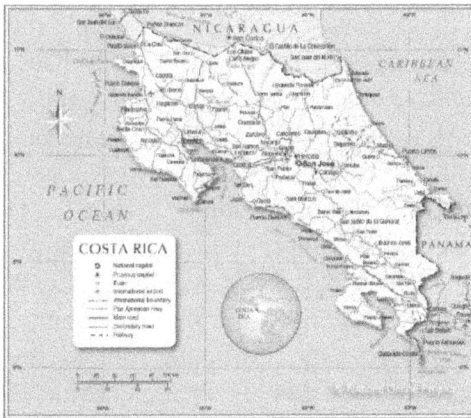

Costa Rica has achieved strong well-being and robust economic growth. The Global Competitiveness Index 2018 rankings placed Costa Rica at 55th position among 140 economies. Costa Rica, and its capital city, San Jose, have been regarded as a center for astonishing ideas and is home for noble innovators. According to the 2019 World Bank Ease of Doing Business report, the country is ranked 67th among 190 economies. Its stable economy, democratic government, and peaceful society, with no need for the military, have earned Costa Rica the nickname "Switzerland of Central America."

Costa Rica is home to many of the world's exciting and innovative startups. Every year, new entrepreneurs enter the market, hoping to comprise as number one company in their field. There are close to 300 startups in the country, most of them in San Jose. BildTek–construction startup; Bellelli–edutech startup; GoPato–messenger service; HuliHealth–MedTech startup; Slidebean–cloud-based software solution; Speratum–biotech company; are some of the most renowned startups.

In Tholons Services Globalization Index 2019, Costa Rica is ranked 28th amongst the Top 50 Digital Nations. Its capital, San Jose is also placed handsomely at 25th position among the Top 100 Super Cities in the same index.

Costa Rica is most suited for startup ecosystems in the locale. Entrepreneurship has picked up noteworthy consideration in Costa Rica in the recent past. SkyLoft, Studio Coworking & Learning, Impactico, Co Spacio, Creasala Coworking Café are creating the new workspaces. 500 Startups, Dreamit Ventures are among the US accelerators present in Costa Rica. Carao Ventures is the most famous venture capital fund in Costa Rica.

Some organizations that have been promoting the development of the entrepreneurship community in Costa Rica, including incubators, accelerators, and dedicated programs: AUGE (incubator), Carao Ventures (VC and accelerator), ParqueTec (accelerator and incubator), Impactico (co-working space), CIE-TEC (incubator), Sistema Banca para el Desarrollo (funding), UNA Emprendedores (incubator). EYCA (consulting), Sfera Legal (law firm), Arias & Muñoz Costa Rica (law firm), and Microsoft BizSpark (program) are some companies and mentors who have had an important role in the ecosystem, such as supporting entrepreneurs in various areas of knowledge.

Technological infrastructure and a highly qualified talent pool are amongst the strongest points of the Costa Rican startup ecosystem. The nation has been able to pull in an increasing number of direct foreign investments and has persuaded multinational companies such as Intel, HP, IBM, VMware, Accenture and others to expose or grow their operations in the region.

Costa Rica might not be the Silicon Valley of Latin America yet but are moving ahead. There's a very educated workforce that is now finding incentives and programs to assist them to innovate and turn their ideas into viable businesses.

Germany

"Technology is a word that describes something that doesn't work yet."
— **Douglas Adams**

Germany is a Western European country, officially called the Federal Republic of Germany. It is the largest country in central Europe and a global leader in innovation. The country has outstanding universities and research institutes including major engineering, manufacturing, and IT industries. Germany consistently ranks among the world's best for scientific research and is known for its precision and high-tech products globally.

Germany is one of the world's main economic powers. It has the fourth-largest nominal GDP in the world based on IMF World Economic Outlook 2018 report. Germany boasting the largest economy in Europe is an economic powerhouse. According to the World Banks Doing Business 2019 report, Germany ranks 24th among 190 countries globally. It has a highly developed social

market economy characterized by a qualified labor force, developed infrastructure, large capital stock, and a high level of innovation.

Germany has over 8,000 active startups and ranks ninth globally among 202 countries, based on the strength of its startup ecosystem. The country has the most vibrating startup boosters in Europe and is known for fostering innovation and talent all across the continent. Capital city Berlin is the startup hub of Germany followed by Hamburg, Munich, and Ruhr. The country has produced 10 of the most successful unicorns: Auto1–online marketplace; Otto Bock HealthCare–prosthetics company; N26–mobile bank application; FlixBus–transportation service; NuComGroup–consumer services; CureVac–Medtech; Celonis–enterprise software; About You–e-commerce; Omio–travel website; GetYourGuide–online marketplace. There are 106 accelerators and incubators, 454 venture capitals and 334 co-working spaces in Germany.

According to the Tholons Services Globalization Index–2019, Germany ranks 35th among top 50 digital nations. German cities also featured in the Top 25 Potential Digital Leaders ranking Berlin (9th) and Munich (22nd) position.

Germany is home to many successful incubators such as Axel Spring Plug & Play, Black Forest accelerator, German Silicon Valley accelerator, Hubraum, Finleap, GTEC, HitFox Group, MAKERS. The BVIZ is the Federal German association of innovation, technology, and business incubation parks launched in 1985. It is one of the most experienced associations of incubators that provides support in building and developing the necessary infrastructure for technology parks and promotes technology transfer, entrepreneurship, and knowledge-based startup firms. German Silicon Valley Accelerator alumni companies have raised USD 150 million in funding and 92% of startups supported through this program are still in business. The startups in Germany get extensive mentoring, coaching and funding from these incubators.

Germany has one of the most innovative business strategies and fresh content models for the entrepreneurial sector. The Top 100 startups in Germany have attracted more than USD 5.9 billion funding since their foundation. The recent few years have seen diversification in investments across various stages of funding. The significant Series A/Seed investors of the country are b-to-v, Blue Yard, Capital, Capnamic Ventures, Cavalry Ventures, Cherry Ventures, Creathor Venture, E. ventures. Whereas, Acton Capital Partners, Global Founders Capital, Index Ventures, Kinnevik, Kite Ventures, Lakestar are majorly investing in Series B. German VCs pumped USD 5.8 billion and 48% of these VCs are backed by international investors.

German science and technology innovation show diversity, by having many EU-funded projects coordinated by German companies, universities, and research institutes. They offer numerous incentives to investors and are formulated to support companies at all stages of the investment process. Promotion of business expansion, health care, new investments, infrastructure, agriculture, and renewable energy are the areas of incentive focus. Most of the investment and operational incentives are provided as reduced tax loans, subsidies, and public grants. The German government has also launched many events and unprecedented campaigns to promote the advancement of new technologies.

Germany's startup scene is paying off with innovation. With strong economy and favorable business microclimate, numerous promising startups are flourishing in the country. The advanced business culture, modern business tools, talented young people, and fresh ideas are strengthening the spirit of entrepreneurship and innovation. With easier access to capital from VC's, favorable tax conditions, and increased recognition of entrepreneurial achievements is not only creating a better environment for startups but also massively strengthening Germany as veritable innovators.

Hong Kong

"Design is not what it looks like and feels like. Design is how it works."
— **Steve Jobs**

Hong Kong is a special administrative region of the Republic of China is located on the southern coast of China. With a population of over 7.4 million, Hong Kong is one of the most densely populated

cities in the world. Hong Kong hosts the fourth-largest concentration of ultra-high-net-worth individuals in the world and ranks 10th among GDP per capita Ranking 2019 report. The territory has become one of the world's most significant financial and commercial centers.

Hong Kong is identified as an alpha+ global city, indicating its influence throughout the world. The city has the highest Financial Development Index score and consistently ranked as the most competitive and freest economic area in the world. Hong Kong has a capitalist mixed service economy with low taxation, minimal government market interference, and an established international financial market. It is the world's 32nd largest economy, with a nominal GDP of approximately HK$ 2.74 trillion (USD 381 billion) with services sectors accounting for more than 90% of GDP.

Hong Kong's startup ecosystem is flourishing, and there continues to be significant growth in the number of startups. According to AngelList, the territory has over 2500 startups employing 9,548 employees across 13,756 workstations. Hong Kong ranked 28th position globally in Cities Global Ranking of Startup Ecosystem 2019 report. With a startup valuation of more than USD 15 billion, Hong Kong is making a name for itself in the global startup world. Hong Kong has managed to produce unicorns far more effectively. The territory has proudly boasted 7 unicorns, Bitmex–trading platform; GoGoVan–transportation and logistics system; Klook–services booking platforms; Lalamove–delivery platform; Sensetime–AI startup; Tink Labs–lifestyle business providing services; WeLab–mobile lending platform operator.

Hong Kong has more than 12 accelerators, 7 incubators, 60 co-working spaces, and 14 venture capitals. Ablaze, Accelerate, Brinc Ltd, Betatron, Cyberport Incubation Programme, Fintech Innovation Lab by Accenture, Hong Kong Design Centre's Design Incubation Programme, Hong Kong Science and Technology Parks Corporation's Incu-App Programme are some of the top incubator and accelerator programs. Hong Kong startups have attracted the attention of global venture capital firms. 500 Startup, Cocoon's Ignite Ventures, Fresco Capital, Mind Fund, MindWorks Ventures, Nest, Mount Parker, Nova Founders Capital, Regatta Capital Management, True Global Ventures, Vector Ventures have presence in Hong Kong.

The government of Hong Kong has initiated many programs to help overseas and set up local startups in Hong Kong. Cyberport, an innovative digital community runs the incubation program to help digital media SME's and startups to implement their ideas into businesses offering facilities like crowdfunding and peer to peer lending. The Monetary Authority with Securities and Futures Commission deal with balancing demands of investors, realizing risks and improvising in the field of development of new products and services. The companies needing finance/equipment will be helped by SME Loan Guarantee Scheme (SGS) with an amount up to HK$ 6 million. The government has also initiated innovation and technology funds to support companies to upgrade technology.

Hong Kong's legal environment and open market added advantage for innovation. Huawei, one of the world's top telecommunication equipment providers, established substantial scientific cooperation with universities in Hong Kong. They have spent millions of dollars in aid to foster Hong Kong's potential as an innovation hub. Hong Kong has been trying to step up their game in the innovation sector. A whopping amount of HK$ 17 billion (USD 2.17 billion) was passed by the financial secretary under spending initiatives for innovative companies. The prime focus was on financial technology or Fintech which is prominent for the growth of startups. Steering Group an advisory committee also associated the funding program and guided steps to advance the entrepreneurial sector and encouraged local talent.

Hong Kong is Asia's most vibrant city and is the ideal place to do business. The excellent infrastructure, pro-business policies and strategic location makes it the preferred destination for businesses large and small. In addition to its timeless economic strengths, the pioneering entrepreneurial spirit of Hong Kong and its boundless energy create the perfect atmosphere to start a business. Enabling this trend are a number of factors including a strong network of incubators, accelerators and co-work spaces, deep and diverse community partner support, pro-innovation policies and a multicultural ecosystem. It is no surprise that Hong Kong is one of Asia's most dynamic startup destination.

India

"The last mile of the digital highway is not infrastructure but skills of the users."
— Debjani Ghosh, President, NASSCOM

India officially the Republic of India is the largest country located in South Asia. It is the second-most populous country and the seventh-largest country in the world. India has consistently been one of the fastest-growing economies across the globe and the most comprehensive outsourcing destination for both onshore and offshore services in the IT industry. It is all set to benefit from the emerging IT skill and talent capitalizing them on its digital technologies.

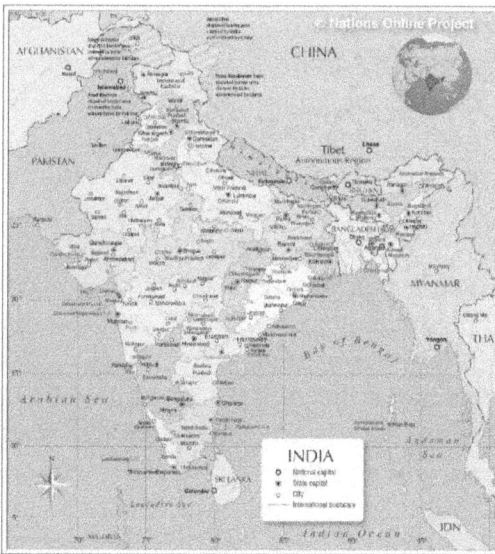

India is a developing mixed economy. It is ranked the world's seventh-largest economy by nominal GDP and the third largest by purchasing power parity. Services, agriculture, and industries are the three major sectors according to the nations GDP. The services sector has attracted significant foreign investment inflows and contributes to mass employment and exports opportunities. As per the IMD World Competitiveness Ranking 2019, India is ranked 43rd place globally. The success of the ongoing market reforms in the country is reflected in its competitiveness ranking.

India boasts of being the third-largest startup ecosystem globally, in terms of the number of startups following the US and Britain. According to AngelList, the country has over 47,000 tech startups in diverse industry segments. StartupBlink report 2019 ranked Indian startup ecosystem 37th position among 100 countries. Bangalore is the startup and innovation hub of India followed by Mumbai and capital city Delhi.

Compared to other global markets, India has a modest number of unicorns. 25 Indian startups have evolved to become unicorns: PayTM–digital payment system; Flipkart–online shopping store; Ola Cabs–transportation network; One97 Communications–mobile-internet company; Snapdeal–e-commerce company; BYJU'S–e-learning app; Oyo Rooms–budget lodging website; Swiggy–food ordering application; Udaan–B2B trade platform; Zomato–restaurant search service; ReNePower–renewable energy market; BillDesk–online payment gateway; Delhivery–logistic courier service; Hike–online messaging platform; ShopClues–online marketplace; InMobi–mobile advertising services; PolicyBazaar–insurance aggregator website; Dream11–sports gaming platform; Bigbasket–online supermarket; Rivigo–logistics company; Quikr–classified advertising platform; Icertis–contract management; Lenskart.com–online optical retailer; Freshworks–business support service; Druva–software service solution are the Indian startups that have stepped into the billion-dollar unicorn club.

Tholons Services Globalization Index 2019 ranks India first among the Top 50 Digital Nations. *Silicon city* Bangalore has been the undisputed leader in services globalization for over a decade and ranks first among the Top 100 Super Cities. The report also ranks 10 other Indian cities: Mumbai (4), Delhi (7), Hyderabad (8), Chennai (13), Pune (16), Kolkata (41), Chandigarh (44), Coimbatore (48), Jaipur (50), Ahmedabad (56) in super cities ranking. There are about 274 accelerators and incubators, 379 venture capital funds and 1024 co-working spaces in India.

The huge success of the Indian startup ecosystem can be linked to the continuous and engaging support the Indian government has provided through various schemes and tax incentives. They have introduced more than 50 initiatives to support startup growth in the past few years. Many of these initiatives like "Fund of Funds" and tax exemptions gained hype across the Indian startup community. "Startup India" is a flagship initiative of the Government of India aiming to build a solid startup and innovation environment that helps startup growth enabling continuous economic development and increased employment opportunities at a large scale. The initiative promotes industry-academia partnership and grants funding support to new tech startups fueling the growth of India's entrepreneurial sector. The program also provides specialized funds, international patent protection, venture capital investment schemes, raw material assistance, single point registration, and many other regulatory incentives.[76]

Indian startups are also seeing support and growth opportunities from non-government business incubators. These incubators provide financial and operational support to startups at all stages of development. Centre for Innovation Incubation and Entrepreneurship IIM Ahmedabad, Indian Angel Network (IAN) Incubator, Nadathur S Raghavan Centre for Entrepreneurial Learning (NSRCEL), IIM Bangalore, and Angel Prime are some of the major business incubators in India. To encourage entrepreneurship, the government of India has started venture capital funds which provide capital and strategic inputs. SIDBI Venture Capital Limited (SVCL) and IFCI Venture Capital Funds Ltd. (IFCI Venture) are major venture capital funds controlled by the central government of India. There are many regional venture capital firms like Hyderabad Information Technology Venture Enterprises Limited (HITVEL), Kerala Venture Capital Fund Private Limited controlled by state governments. Indian startups have attracted the attention of some of the globally renowned venture funds. Helion Venture Partners, Accel Partners, Blume Ventures, Sequoia Capital India, Nexus Venture Partners, and IDG Ventures, etc. have their proximity in India.

India is the undisputed leader in tech and BPM for over two decades. It is emerging as the leading location for digital skills and solutions to multinational corporations. The country has a vibrant startup ecosystem and is evident from the growing number of unicorns the country is producing. India's rapid growth of digital skill talent and blooming startup scene showcases its tremendous potential to outperform the technology leaders in the future.

Ireland

"Ideas are commodity. Execution of them is not."
— **Michael Dell**

Ireland is an island in the North Atlantic Ocean, located in the north-west of Europe. The population of Ireland is about 6.6 million, ranking it the second-most populous island in Europe. The Country is divided between the Republic of Ireland (officially Ireland) covering five-sixths of the island, and Northern Ireland, which is part of the United Kingdom. Despite the two jurisdictions using two distinct currencies (the euro and pound sterling), a growing amount of commercial activity is carried out in the country.

The economy of Ireland is fundamentally a knowledge economy, focused on services in high-tech, life sciences, financial services, agribusiness, forestry, fishing and mining. Ireland ranks 5th in the IMF table of GDP (Nominal) per capita Ranking 2019. It is an open economy and ranks first for high-value foreign direct investment (FDI) flows. The World Banks Ease of Doing Business Index 2019 report places Ireland at 23rd position.

According to Startup Ecosystem Rankings 2019, Ireland is ranked 14th position globally. Dublin and Cork are the major startup hubs of Italy. Forbes named capital city Dublin as "The best city to finding a business in." Most of the Global Tech companies such as Facebook, Google, LinkedIn and Twitter who rank in the top 10 are in Dublin, making it a global tech hub. These serve as pillars in ranking Dublin as 8th in the European Digital City Index (EDCI).

In Tholons Services Globalization Index (TSGI) 2019, Ireland is placed at 14th position amongst the Top 50 Digital Nations, due to its active outsourcing industry coupled with the booming startup scene. Amongst the Top 100 Super Cities, Capital city Dublin is also placed at 5th position and Cork, the second-largest city is placed at 62nd position.

Ireland boasts over 2500 startups, 165 hubs, and 250 accelerators. Some of the renowned startups of Ireland are: Cortechs–a brainwave-sensing headset that reads brainwaves and regulates players' focus level; Jobbio–job search platform; Drop–Foodtech; Mobacar–mobile-based CAR automation system; Coras–Ticketing startup; Yroo–search engine for smart shoppers.

To empower entrepreneurial teams in transforming strong ideas into commercially possible startups, Ireland started its first digital accelerator–National Digital Research Centre (NDRC). It was found in 2007 and invests in companies having the potential to grow internationally. It has worked with over 200 startups. NDRC-supported companies have raised over USD 166.7 million and employed over 800 people. As a University Business Accelerator, NDRC is currently ranked second in the world and first in Europe by the UBI Global Index.

In making Ireland as the best European startup hub, the government and agencies are investing a lot of resources. The global program, 'SAP Startup Focus program' is helping startups in the big data, predictive and real-time analytics space. The program's encouragement in developing new applications on SAP HANA is accelerating market traction. Multitudes of incubators and accelerators are active in Ireland startups. The National College of Ireland's Business Incubation Centre, University College Dublin's Nova and Invent, and DIT Hothouse provided by Dublin Institute of Technology are the other leading university incubators supporting the startup ecosystem.

There are many grants which make sure no genuine idea is left behind for lack of finances. For example, Dublin Startup Community Fund, which is provided by the Dublin Commissioner for Startups, gives small grants to community-focused entrepreneurs and organizations in Dublin. IBYE (Ireland's Best Young Entrepreneur Competition) is a program run by the 31 Local Enterprise Offices (LEOs) with the support of the Department of Business, Enterprise and Innovation. The program founded in 2015 have funded over USD 7.2 million to 500+ young entrepreneurs.[77]

Ireland has many Angel and Seed Investors. Prominent among them being Halo Business Angel Network, Irish Investment Network, Business Angel Partnership, AngelList, Lucey Fund, and AIB Seed Capital Fund provided by Dublin Business Innovation Centre. Enterprise Ireland (EI) has been ranked third in the world for seed investment by the international investment platform, PitchBook. Startupbootcamp Ireland, Google's Ventures fund and Mastercard's e-commerce-focused accelerator, are among the other sources of funding in Ireland.

The presence of many international and local venture capital investors is a big positive for Ireland. NDRC VentureLab Program, ACT Venture Capital, Frontline Ventures, Enterprise Equity Venture Capital Group, Delta Partners, Kernel Capital, Business Venture Partners, Atlantic Bridge LP, DFJ Esprit, Executive Venture Partners, Fountain Healthcare Partners, Greencoat Capital, Investec Ventures, MML Growth Capital Partners, Polaris Partners, Serbia Kernel, Irish Venture Capital Funds, ScaleFront, Smart Start, and Irish Venture Capital Association are local VC firms.

Ireland is the place to be for B2B businesses. There are a large number of companies in Ireland looking for new ideas, tools, products, and services to help them grow and improve. However, no unicorn has come out of Ireland yet and there is also limited startup growth outside Dublin. Ireland has more resources to encourage the creation, scalability, and stability of startups. The future of startups looks certainly very bright for Ireland.

Israel

"Let's go invent tomorrow instead of worrying about what happened yesterday."
— **Steve Jobs**

Israel, a country in Western Asia is located at the eastern end of the Mediterranean Sea, and the northern shore of the Red Sea. Israel has marked its rank in economic and industrial development as one of the most advanced country. The credentials of quality education system and reputed universities have contributed in establishment of a highly motivated and educated populace. This has encouraged country's rapid economic development and high technology boom.

Israeli Tech sector is an unprecedented success story, enjoying a period of consistent growth over multiple decades. With Israeli organizations spearheading propelled arrangements in fields including digital health, cybersecurity, fintech, mobility, smart transportation and innovation, Israeli inventiveness is at the height of worldwide development. Israel ranks first in the world in R&D

investment contributing 4.3 percentage of its GDP. According to a new report, Israel's development of cutting-edge technologies in software, communications and, the life sciences have evoked comparisons with Silicon Valley. The country is ranked 20th in the World Economic Forum's Global Competitiveness Report 2018 and 49th on the World Bank's Ease of Doing Business index.

Israel with over 5,000 startups is an innovation giant. Two venture capital-funded startups have made entry to the unicorn list. The country ranks 4th place in startup-ecosystem-rankings-report-2019. In many ways, the strong culture of Israel's creative innovation runs parallel to that of America. Both countries share the unique view that business failure is not a tag of dishonor but an opportunity. They do not prosecute risk-taking.

According to Tholons Services Globalization Index (TSGI) 2019, Israel ranks 41st among the Digital Nations. Tel Aviv ranks 71st in the Super Cities list. In the Startup Ecosystem Index 2019, capital city Tel Aviv ranks second globally. Although the country is not having a matured outsourcing industry, it has an established startup ecosystem. Beer Sheva and Haifa also have promising and well-functioning ecosystem for startups and entrepreneurship. One of the most popular and widely known Israeli startup is Waze, a mobile phone traffic and navigation app that has created a "driving community" and is used by nearly 100 million drivers worldwide. Waze has been so successful that it was acquired by Google. OrCam developed technology which allows blind people with intact optic nerves to see. Phinergy's lithium car battery triples the mileage of electric cars. These are only few examples of numerous such startups making Israel a "startup nation."

Israel boasts of having the greatest number of startups per capita and attracting more venture capital dollars per capita than any other country, hence the nickname 'The Startup Nation'. With a population of around 9.5 million, it has the largest number of startups per capita in the world, around one startup per 1,400 people. Education has played a crucial role in Israel earning that nickname. In a survey done by cloud accounting software firm Sage, Tel Aviv University ranks eighth in the world for training entrepreneurs. It had seven alumni founders of startups with valuation worth USD 1 billion and above.

The Israeli government has been designing and implementing strategies and policies to foster startup ecosystem in Israel for decades. To stimulate seed financing, government has introduced new guidelines wherein personal income tax deduction is provided for amounts up to NIS (New Israeli Shekel) 5 million to those who invest in startups that are less than three years old and have a turnover that is less than NIS 1.5 million. Israel's corporate tax rate of 25% is high compared to international standards. However, the government offers tax breaks to foreign investors, exporters and R&D-based companies through the Law for the Encouragement of Capital Investment.

An average of 600 new startups are established every year in Israel and there are 307 multinational R&D centers, including Intel, Google, IBM, and Apple. There are currently 225 hubs including co-working spaces and entrepreneurship programs in the nation. There are approximately 128 venture capital funds and over 87 incubators across the country, all of which have been privatized. The corporates—Yahoo, Citi Bank, Aol, Microsoft, Upwest Labs, and TheHive have been running accelerators in Israel, and presence of leading venture capital firms has helped the startups raise funds of over USD 5.5 billion in 2017.

The startups in Israel are sources of groundbreaking innovation and are taking Israel to global tech prominence. The combination of Israel's culture, environment and keen strategy has equipped Israel with a startup ecosystem that rivals any other top startup hub. Israel's ability to swiftly translate market demands into organizational action accounts for its consistently strong performance in the flexibility index and its broad acceptance as an innovation capital. The success of startup culture in Israel is an inspiration and a perfect example of how culture, environment and strategy can overcome size limitation.

Italy

"That the way to achieve higher standards of living for all is through science and technology, taking advantage of better tools, methods and organization."
— **Charles E. Wilson**

Italy is a Southern European country officially known as the Italian Republic. With a population of 60 million, the country is divided into 20 regions. Due to its central geographic location in the Mediterranean and Southern Europe, Italy has historically been home to myriad people and culture.

The north of Italy is much dominated by developed industrial economy by private companies; while the south with less-developed agricultural welfare-dependent economy. The northern regions are the "economic engines" for Italy. Food, machinery, iron and steel, footwear, clothing, and ceramics are the main businesses in the northern region. Contrastingly, the prosperity of southern regions is less, and there is a gap in the economy. Southern Italy's economy is dependent on small enterprises, primarily – agricultural, manufacturing and the tourism sector.

Italy has an advanced mixed economy, ranked as the third largest in the Eurozone and the eighth largest in the world. The country has been ranked 50th among 190 countries globally in the World Bank's Doing Business Report 2019. Italy is known for its excellent creativity, extraordinary entrepreneurship and style.

Italy startup ecosystem is ranked 25 globally with more than 3000 startups listed on Tracxn–research partner for Venture Capitalists. Milan and Rome are the two major Italian startup hubs. Milan has all necessary elements to build a successful startup ecosystem, from Venture Capitals, university incubators, accelerators to a variety of startup events. FabriQ, Digital Magics, Make a Cube, and Polihub are some of the top incubators in Milan. Rome, located at the heart of Italy provides all the support required for fostering startups and entrepreneurship. It hosts many entrepreneurial events like Roma Startup, Startup Italia Roma, Eventbrite, etc.

Italy is fast becoming a hotspot for new tech talent despite being a latecomer to the booming startup scene. The renowned startups of Italy are Kaitiaki–digital social ecosystem; Mygrants–online educational platform; Carplanner–car subscription marketplace; Nearit–contextual mobile engagement web platform; Freeda Media–feminist editorial startup; Eatsready–foodtech; Splittypay–fintech; HeartWatch–medtech; Bigprofiles–Customer Intelligence tool. Romanian startup UiPath, building AI-based services for enterprises in the area of robotic process automation has recently joined the unicorn list by reaching a market value of over USD 1 billion. Italian innovative startups have made almost €1 billion (USD 1.11 billion) in total turnover.

In recent years, the number of accelerators and incubators supporting new businesses has grown in Italy. There are more than 50 smart working centers, 12 fab-labs, 20 incubators, and 5 tech-transfer centers. Copernico is an interconnected workspace for professionals, startups, and corporations that promote and grow business. StartMi Up is a startup community created by Marseille Innovation. They aim to create smart working environments that connect professionals from different, yet related fields with private and institutional investors, to generate new ideas, projects, and enterprise.

The number of private, public and international initiatives to sustain innovative entrepreneurship in the territory is exploding. There are several university incubators like Bocconi University incubator and the Politecnico di Milano incubator also called Polimi. 'Startcup Lombardia' is a competition organized by the Universities and University Incubators favoring the creation of innovative potential businesses. Eppela, SiamoSoci, and Starteed are the major crowdsourcing platforms. Innogest SGR, 360 Capital Partners, Dpixel, KeyCapital, and United Ventures are the major venture capital firms of the city.

Italy provides central government grant, redeployment grants, European Union (EU) subsidies, and grants from provincial authorities and local communities to foster innovation. The Italian Startup Act launched in 2012, made it simpler to register a new company, relaxed labor regulations, introduced tax incentives for investors and created a fast-track startup visa for entrepreneurs. The Act provides tools at all stages of the business life cycle, creating the enabling conditions needed for a quick go-to-market and scaling up of high-tech startups. Under this act, innovative startups can get several numbers of incentives for the first five years. These include tax incentives by up to 27%, free-of-charge access to public guarantees by 80% on bank loans amounting up to 2.5 million Euros.

Italy has a rich history in innovation, creativity and entrepreneurship It is a country of innovators, entrepreneurs and risk takers. The growth of the startup scene in Italy has gained some pace which a reflection of regulatory reforms and initiatives is taken by the government to support innovation and entrepreneurship. Despite these gains, investment opportunities in Italy remain low. Perfecting the existing policy architecture and utilizing the VC's and government reforms might indeed help build on emerging momentum and further nourish the nascent startup ecosystem.

Japan

"Any sufficiently advanced technology is indistinguishable from magic."
— **Arthur C. Clarke**

Japan is an island country located in the Pacific Ocean lies off the eastern coast of the Asian continent. It has a total of 6,852 islands extending along the Pacific coast. Japan is the third largest national economy in the world in terms of nominal GDP. It has a large industrial capacity and is home to some of the largest and most technologically advanced producers of motor vehicles, electronics, and machine tools.

Japan ranks 39th of 190 countries in the World Bank Groups 'Doing Business 2019 report'. Japan has played leading role in designing the technology landscape. Japanese companies like Toshiba and Sony make an interesting comparison with Silicon Valley's recent technological developments.

The large corporations of Japan have a significant hold and influence over the market and talent pool. Silicon Valley-type startups like Amazon, Uber, and Tesla that are changing lives and economy of nations like USA are to be witnessed in Japan. These startups have the potential to reform the present industries and economy. Despite having a reliable talent pool, capital, infrastructure, and technology readiness, the least risk-taking culture of Japan hinders its startup growth. In Japan, the innovation and corporate works on step-by-step, predictable and low-risk model, while Silicon Valley's strength is its "fail fast, fail smart" model of innovation.

According to the Global Competitiveness Report 2018, Japan has been ranked 5th out of 140 economies, considering its higher level of innovation, business sophistication, market size, and infrastructure. As per *Innovation Cities™* Index 2019, capital city Tokyo ranks first in the list. Osaka (45th), and Kyoto (64th) are named Top 100 global cities for an innovation economy. Japan wants to formally encourage domestic startups to pursue growing opportunities in commercial space. The Japanese government has earmarked around USD 1 billion in public funds to startups working on space solutions[78]. The local municipalities launched the Startup City Promotion Consortium in cooperation with economic organizations to promote entrepreneurship.

The National Strategic Special Zones were designed to elevate the global competitiveness of the industry and promote international economic activity centers. The government of Japan has started several programs to support startups. To name a few— Sido Next Innovator by Ministry of Economy Trade and Industry; BusiNest Accelerator Course and Seed Accelerator Program (Osaka prefecture, Initiative for Development of Entrepreneurship in Asia). Several pitch sessions and competitions are being organized throughout the year in the country, such as the Samurai Venture Summit, Tokyo Startup Gateway, Innovation Weekend, Infinity Ventures Summit, SF Japan Night, and Decoded Fashion Startup Summit to connect entrepreneurs and investors.

As per AngelList, Japan has more than 1200 listed startups. It has 10 unicorns that include Rakuten and Dena–e-commerce companies; Cookpad–food tech company; CyberAgent, and GREE– provision of Internet media services; M3 Inc.–medical web portal; Mixi–social networking service; GungHo, Colopl and Kakaku– smartphone application provider. Mercari, an e-commerce company also joined to the nation's unicorn list in 2018. The Japanese government is aiming to produce 20 unicorns by 2023. The Japanese Ministry of Economy Trade and Industry (METI) have launched the startup support program named 'J-Startup', aiming to promote overseas development of Japanese startups and creation of unicorns. The program chooses 100 startups and certifies them based on their performance and expertise.

Japan has regulated several policies to catalyze startup ecosystem growth. The strategies include—providing education and information to entrepreneurs through National Startup and Venture Forum; reducing the minimum capital needs for starting limited liability company, a new startup loan program that doesn't require collateral/guarantors.

Japan has over 47 business incubators. Some of these are focused on the hi-tech sector digital development–offering funding, workspace and mentoring. Japan's large corporations are typically conservative and domestic in their business strategy. However, a major change has been taking place over the past few years. Corporations are trailing investment opportunities with startups and business development, sensing the limits of developing new services and products in-house.

London

"The great growling engine of Change-Technology."
— **Alvin Toffler**

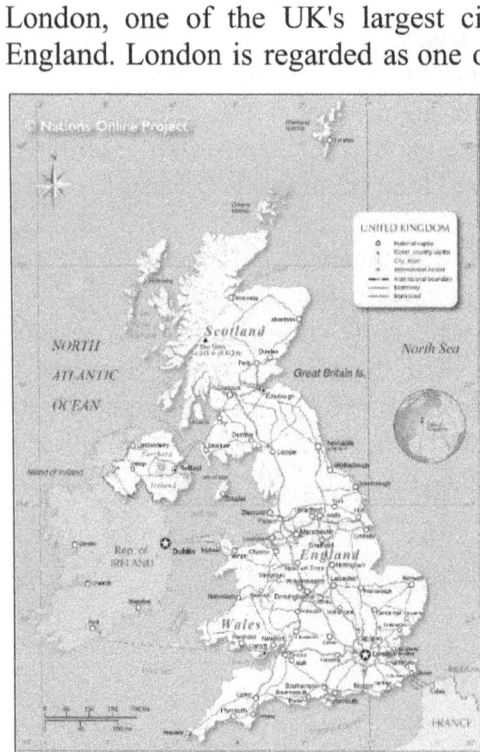

London, one of the UK's largest cities and the capital city of England. London is regarded as one of the most significant global cities in the world and has been named the strongest, most attractive, most influential, most innovative, most competitive, most investment-friendly city in the world. It has a significant effect on trade, schooling, fashion, finance, healthcare, media, skilled facilities, and R&D. It is known as one of the biggest economic centers with the highest metropolitan GDP region. It is one of the leading investment destinations for international retailers and ultra-high-net value individuals than any other city.

London solidly built up its notoriety as the Fintech capital of the world. In Tholons Services Globalization Index 2019, London ranks 27 amongst the Top 100 Super Cities. Investors craving for troublesome new businesses in the monetary administrations advertise has expanded beyond Angel and Seed financing rounds to genuine multi-million-dollar Series A and B venture rounds. Health Tech or Digital Health is also quickly picking up, consideration from the "startup ecosystem" of business visionaries, government activities, corporate hatcheries, angel speculators, and venture capitalists in London. Money transfer, foreign exchange, and payments-focused tech companies are progressing fast and situating themselves as market leaders in multi-billion-dollar untapped markets with the potential for exponential development.

The success of the revolution in London has brought in the unbundling of other industries as well. Digital jobs are on the rise with over 250,000 such jobs and expected to rise by more than 45,000 in the upcoming years. Over 16,500 startups are active in the London Metropolitan Area. Sunlight–a smart digital wallet and marketplace for employee learning; Mainframe–a decentralized communication layer for the new web; Gojimo–Exam preparation platform; Daisie–creative collaboration platform; Hackajob–a job application platform; Peak–a brain training app, are some of the few startups in London.

There are 13 startups being valued at USD 1 billion and above, that are BenevolentAI–AI-based medical startup to develop new medicines for neural disease; Checkout.com–a leading international payment gateway; DarkTrace–a security company to fight against cybercrime; Deliveroo–online food ordering and delivering app; Graphcore–machine learning and artificial intelligence-based processor developer; Immunocore–T Cell Receptor (TCR) biotechnology company that delivers first-in-class biotherapy to cure serious illnesses; Improbable–a gaming company; Revolut–a UK financial technology company that offers banking services in currency exchange (money withdrawals in 120 currencies and sending in 29 currencies), cryptocurrency exchange and peer-to-peer payment through the app. It also provides customers access to cryptocurrencies; Monzo–a digitalized mobile bank app; OakNorth–small and medium loan lenders; Oxford Nanopore–DNA and RNA sequencing technology; The Hut group–health, fitness and lifestyle market firms; Funding circle–a peer-to-peer finance lender; Transferwise–a fintech company; BrewDog–a multinational brewery and pub chain company.

London is adjudicated as having the best startup hubs in Europe. There are over 150 coworking spaces present in London such as We Work, Google Campus, The Office Group, Rainmaking Loft, etc. It has a presence of 360 incubators and accelerators, including Europe's largest fintech accelerator Level39, helping youthful business visionaries and cultivating a startup environment. Health-focused startup "RotaGeek" was shortlisted for Wayra, UK's 2014 mentoring program. National Health Service (NHS)-focused startup Healthberry beta launched to measure patient feedback. Other companies like Visa are looking into London's potential as an innovation hub accelerating resource. Innovators in the payments space can apply to be part of Visa Europe Collab, the company's innovation program in Shoreditch. In its first year, the company points to take at least 20 ventures through its 100-day development funnel, from starting scoping and capability, through advertising testing, design, and confirmation of concept. The selected ones be sent to the main Visa business to be used by its consumers, retailers and member banks across Europe.

The British government is actively supporting the startup ecosystem by providing £ 25,000 (USD 32K) at a fixed interest rate of 6% per annum to new businesses. This scheme known as Start Up Loans was started in 2012 and has pumped more than £ 500 million (USD 642 million) in over 63,000 business ideas. Innovate UK, a government body, holds funding competitions for businesses and research organizations. Tech.London, a joint project between the Mayor of London, investor portal Gust, and IBM, provides resources to startups such as tips on setting up, workspaces and events to mentorship programs, jobs portal, and funding tips.

Even universities and colleges are taking a bite of the innovation progress in London. The I-HUB, Imperial's new translation and innovation hub at the White City Campus, gives modern incubator, accelerator, office space and laboratory, which is more than 185,000 square feet. the facility empowers the co-location of startups, business visionaries, and major enterprises nearby Imperial's academic community.

London, being a major metropolitan city, attracts talent and resources from around the world. Investors are tapping into this gold mine, only to discover that the possibilities are endless. With this momentum and success, London might soon become a more thriving technological hub.

New York

"It's not that we use technology, we live technology."
— **Godfrey Reggio**

New York is the strongest financial location and it is one of the world's best tech innovation hubs. Majorly tech companies are being instrumental in generating an ongoing boom in innovation for consumers, employers and merchants.

New York has been ranked first in the Savills tech cities

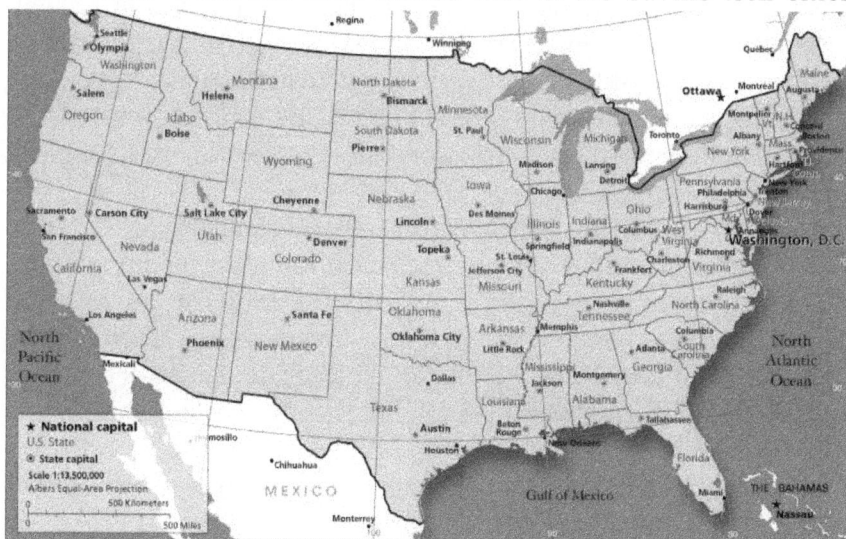

index and second in the 2019 Top 20 Global startup ecosystems list. In Tholons Services Globalization Index 2019, it is ranked 29th amongst the Top 100 Super Cities.

New York has over 28,950 startups in AngelList. Some of the renowned startups in the list include: themednet–an online community of doctors to facilitate the clinical solutions between the physicians; One Month–a technology company that focus on self-paced online learning; Giphy–online database that allows users to search and share short looping videos with no sound, that resemble animated GIF files; Morty–online platform that allows people shop, lend or compare on property and various other. New York City has startup unicorns like Tumblr, AppNexus, MongoDB, Buddy Media, Gilt Groupe, Warby Parker, Shutterstock, Kickstarter, Gerson Lehrman, Etsy, and OnDeck Capital.

Startup 51 maps provide safe platform that allows companies to merge all the applications utilized by employees into one central secure dashboard that can be accessed on any gadget. AlphaPoint is another firm that provides blockchain-enabled solutions to digital assets being issued, tracked, and traded. ForwardLane offers benefits that utilize artificial intelligence to enhance the essence and sophistication of investment advice. The wealth management solution enables financial advisors to render personalized, high quality and differentiated services to clients. These are just a few instances of technological innovation that New York has to offer as a hub. The world of financial services is demanding innovative ways of accomplishing business. New York is the root of some of the biggest economic traffic in the world, it is innovating to its highest capacity.

New York City has a wide range of company accelerators that provides free office space, seed cash in return for investment, mentors, prospective partners/investors, free legal advice and banking. Entrepreneurs Roundtable Accelerator, Tech Stars, DreamIt Ventures, Runway Startup Postdoc at Jacobs Technion-Cornell Institute, and New York Digital Health Accelerator are some of New York's top tech accelerators. "Rise New York" is a British multinational banking joint undertaking and Barclays is a virtual global community and physical space outlined to lead the economic technology future. Barclays' first Accelerator program "Rise New York" launched in 2014 in London, New York and Tel Aviv with over 140 companies in its portfolio. The program covers all Fintech fields from cybersecurity and artificial intelligence to wealth management, big data, and cryptocurrencies. It incorporates consultancy and opportunities for financial technology startups to approach influencers, industry experts and promising clients. *"The Fintech community is growing faster than ever with a number of game-changing startups. With the expansion to create our largest-ever Rise site, not only can we give those in our Accelerator program the room to grow-but we can also help to house more of the best and brightest innovators."—John Stecher, Group CTO and Chief Innovation Officer, Barclays.*

The influx of business incubators is one of the main factors for enabling New York as a tech center. There are over 70 incubators and accelerators in New York. The companies that have raised the largest startup funding capitals in New York City in Q1 of 2019 are Rent the Runway (USD 125 million)–a subscription fashion service that allows women to rest unlimited data, Andela (USD 100 million)–a software-based company that identifies and supports software developers, Glossier (USD 100 million)–an e-commerce website for makeup products, Casper (USD 100 million)–an e-commerce site that sells sleep products, and various other.

New York universities also invest millions of dollars to foster entrepreneurship and innovation. A team of startup professionals from New York University offers educational training and activities, resources, and funding to inspire, connect, educate and expand entrepreneurs across NYC. STARTUP NY program offers new and expanding businesses to operate tax-free for ten years in New York State. The New York Fintech Innovation Lab is a 12-week highly competitive program that promotes technology businesses in the phase of "Early to development." The New York Fintech Innovation Lab received USD 1 billion in 2019. As of 2019, the laboratory has developed 1100+ employment, with 69 businesses, 44 associates and 241 evidence of ideas.

With the advent of these startups, New York City is innovating financial technology at the grassroots level in Entrepreneurs; it will only be a matter of time when financial technology in New York would come to par or even exceed the pace at which the technological revolution is taking place.

Peru

"Coming together is a beginning; keeping together is progress; working together is success."
— **Henry Ford**

Peru, the third-largest tropical country of South America with over 798,000 square miles of land is referred to as "Land of Abundance."

It is also the fifth most populous country with 32.74 million people, with 79.2% of which are urbanized. Lima, the capital city, is part of America's largest cities, holding 31.7% of the country's population. Lima valley has the presence of an active team of entrepreneurs and because of this, Peru has been considered as an active hub of entrepreneurship in recent times.

In Tholons Services Globalization Index 2019, Peru is ranked as 42nd amongst the Top 50 Digital Nations. Lima is also positioned at 91st amongst the Top 100 Super Cities.

Peru introduced the StartUp Peru (SUP) program in 2012. Since then, it has improved the program design and it has increased the budget for startups, which receive resources from the innovation fund. The StartUp Peru program has been expanded, and it includes seed capital and support for angel-investor networks. Peru also promotes the funding of startups based on scientific research through a competition organized by the National Science and Technology Council (CONCYTEC). Some of renowned startups of the nation list: DevCode–Mobile Application; Jionnus–Event gathering platform; Get onboard–Job application platform; Turismoi.pe–

tourism portal; Crehana–an on-demand learning platform for creative and digital professionals.

In addition to offering direct support to startups, SUP also provides funds for developing incubators and accelerators. Peru Emprende comprises public and private institutions that organize activities and events to strengthen the entrepreneurial scene. It has been active since 2014. WorldLoop, an international non-profit organization that provides support and advice to social entrepreneurs to eliminate the negative impacts of e-waste and Oksigen Lab, a social incubator that provides coaching to Social Entrepreneurs to perform research on Social Entrepreneurship and Innovation.

The Peru Venture Capital Conference is the largest entrepreneurial capital event in Peru that connects the Peruvian innovation and entrepreneurship ecosystem with local and international investors. UTEC Ventures, a business accelerator that is part of UTEC's entrepreneurial department, is one of the most respected organizations in Peru. After beginning in 2014, the UTEC Ventures incubator and accelerator has over 50 businesses in its portfolio and frequently organizes educational activities for learners to educate about startups and innovation. Wayra Peru, a corporate accelerator by Telefonica launched in 2017, provides up to USD 50,000 in funding, access to an accelerator program in Lima, and access to a wide network of global partners, mentors, and experts. Angel Ventures Peru, PAD, BBCS, and Capital Zero are some of the angel networks present in the country. Some of the prominent accelerators are Wayra, Emprende UP, Fledge, and Endeavor.

The ecosystem is ripe for disrupting the incumbent players and Peru is going at a high pace to ensure it doesn't miss this opportunity.

The Philippines

"The betterment of the next generation will be enabled not by politicians, but by engineers and entrepreneurs rich in ideas for innovation, and by businessmen who know the dynamics of wealth accumulation."

— Banatao Founder Philippine Development Foundation (PhilDev)

The Philippines is the 5th largest island country in the world that is rapidly growing. The Philippines is an archipelago of more than 7,000 islands with a total area of 300,000 square kilometers. The Philippines is undoubtedly the leading client service provider catering Business Process Management (BPM), multinational companies (MNCs) and Global In-house Center (GICs). The Philippines' culture, tradition, and English fluency are similar to the US and UK that has been the key driver for its growth and outsourcing opportunity especially the MNC operations.

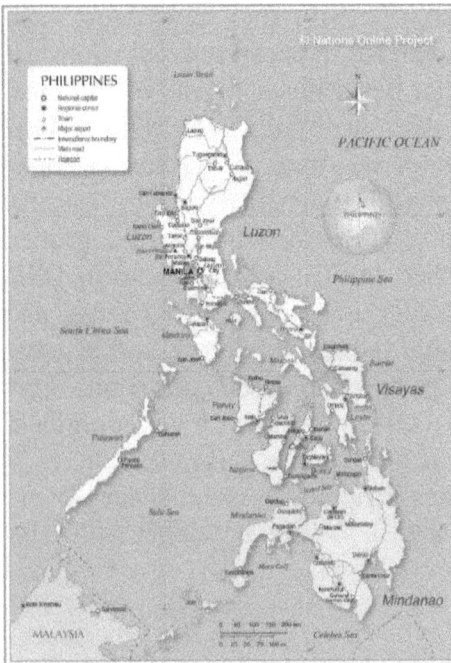

Tholons Services Globalization Index 2019 ranks the Philippines fifth among Top 50 'Digital Nations'. Manila (2nd), Cebu City (12th), and Davao City (95th) are in the Top 100 'Super Cities'. The city of Manila has become the center of operations for various sectors, challenging Bangalore, Mumbai by taking the lead in BPM. The Philippines has a strong youth population. Besides, the country has a substantially high literacy rate (96.5%); an estimated 44% of the graduates are employable and is one of the world's largest English-speaking nations.

The roadmap was intended by the Philippine government and startup community to build a flourishing ecosystem. The Philippines has over 1100 startups in AngelList. Kalibrr–An end-to-end recruitment platform that uses AI to match the recruiter's profile; Talkpush–Automation software for recruitment that allows employers to engage in discussions and interact with thousands of applicants; CompareAsiaGroup–A leading financial aggregator website that compares banking, insurance, and telco products, enabling consumer to find the best products as per the needs; Playlab–Mobile Puzzle Games developers for iOS and Android; PicLyf (pik+life)–A social platform for publishing picture diaries that allow users to create a visual narrative of their Life, Things, and Interests. Revolution Precrafted, the Philippines' first startup is evaluated at over USD 1 billion, is a luxury prefabricated home.

Various incubator programs such as AIM-Dado Banatao Incubator, launched by the Philippine Development Foundation (PhilDev) in collaboration with the Asian Management Institute (AIM), provide mentoring, training, and support network building to grow businesses. Cerebro Labs, a tech business incubator and accelerator for early-stage startups; Ideaspace Foundation Inc, Incubator and accelerator program that support ideas to be developed in commercial product through competition and community advocacy program; Impact Hub Manila is an incubator and co-working network that offers various programs for both local and global opportunities through 6-8 week incubation program; QBO Innovation Hub is a first public-private partnership (PPP) designed to develop expand and scale a sustainable start-up ecosystem in the Philippines; Startup Village is an incubator-accelerator newbie that supports small and medium-sized enterprises. SPrInG.PH, xchange, Launchgarage, A Space's Passport, Sunnyvale Accelerator Plug and Play Tech Center are among the major accelerators and incubators of the Philippines that are a catalyst to hundreds of companies in the region.

There are multiple funding options available for new ventures in the Philippines. Fico Bank, BangkoKabayan Inc.,

Cantilan Bank Inc., and 1st Valley Bank are offering loans and microfinance options to promote startups. IMJ Investment Partners, Tallwood Venture Capital, First Asia Venture Capital Inc., and Investment & Capital Corporation of the Philippines are venture firms that are operating in the Philippines for additional promotion. Manila-based Asian Development Bank (ADB) offered an initial USD 3.6 million grands for the development of inclusive business in the Philippines and the rest of Asia.

The Philippine Congress passed the Innovative Startup Act that supports startups with 10 billion pesos (USD 192 million) venture fund, tax breaks and easier processing of business permits and visas for foreigners. Incentives include tax incentives, including a six-year income tax exemption from the beginning of the company's pioneering business activities, as well as a four-year non-pioneer income tax exemption. Enterprises registered with the Philippine Economic Zone Authority (PEZA) are permitted with additional incentives to assist employers in non-urban areas.

The University of De salle, Ateneo De Manila University, The Asian Institute of Management, and San Beda College include some of the best universities that offer entrepreneurship programs in the nation. The government creates many investment opportunities for both local and foreign investors. To use the incentives, foreign investors must register with the Board of Investments (BOI) under the OIC.

The Philippines has retained its position in the outsourcing industry for years. In recent years, the tech startup has blossomed around the world and the outsourcing industry has been impacted by unique technologies such as automation and artificial intelligence. Filipino outsourcing rivals like China and India have aligned their efforts to achieve global appreciation in technology and entrepreneurship. The Philippine government should capitalize on its technically skilled labor and infrastructure to establish itself as a startup nation.

Poland

"If you think you can do a thing or think you can't do a thing, you're right."
— **Henry Ford**

The Polish economy has encountered an enormous transformation in the past years. Collaborating with the European Union (EU) was another huge leap. Among the startup hubs in Europe, Poland gains a significant name. The BPO sector employing over 270K employees, support some of the leading global companies. The startup ecosystem in Poland is flourishing with promising startups that are emerging across the nation. The startups from Warsaw, Krakow, Pozan, Wroclaw, Gydnia, and Katowice have a large spectrum of industries from health tech to recruitment, machine learning, space tech, facial recognition, and digital authentication. The techniques provided by these startups are robust and show a powerful capacity for international expansion beyond the polish economy.

In Tholons Services Globalization Index (TSGI) 2019, Poland is ranked at a highly competitive 15th position amongst the Top 50 Digital Nations. Three Polish cities namely Krakow, Warsaw, and Wroclaw features in Top 100 Super Cities, with Krakow at 11th position. According to AngelList, the number of startups in Poland is amounted to 1782. Azimo–an online international money transfer company; Contellio–a design automation company that creates infographics and presentation; Brainly–a multinational educational startup that operates a group of social learning networks for students and educators; Gaming Live– a broadcasting service that focuses on European market and Video games as entertainment; inFakt, Niania.pl, Qrcao are few other startups in Poland. Allegro, an e-commerce base is a unicorn; SalesManago and Growbots are the potential "to be" unicorns.

Warsaw is the capital city of Poland and is part of the TWIST Digital project, that connects three other European cities (Rome, Lille, and Stockholm) under the Startup Europe umbrella. Poland brought in € 7.2 billion (USD 8 billion) in EU funds that aimed at promoting an innovative economy after joining the EU in 2014. About 30% of all Polish startups surveyed in Startup Poland's annual report are based in Warsaw. According to Startup Poland, nearly 60% of Polish startups are bootstrapped from their funding. Wroclaw is another important startup hub in Poland with a strong economic structure, providing numerous business opportunities. It gave birth to startups like LiveChat and Cloud Your Car. The country's most advanced tech park resides in Wroclaw. It has attracted the younger generation from nearby regions. The number of innovative and creative hubs, accelerators, and workshops has just added flavor to the city's improvements.

Advancements in Fintech have proved to be a catalyst for many startups. Golem is a decentralized global market, often coined as the "Airbnb" for computers. It accepts payments in cryptocurrency. It allows users to lend resources to people in need. Forbes explained that Golem serves as what many views to be "the backbone" of a decentralized market for computing power, making it both an Infrastructure-as-a-Service (IaaS) and Platform-as-a-Service (PaaS) platform. A complex marketing automation platform, SalesManago, founded in 2011, has raised USD 7.7 million and is now present in over 40 markets. Estimote and Kontakt.io specialize in beacon technology and in bringing the Internet of Things to interactive marketing.

Founder Institute Warsaw, Aip preinkubacja, Startup Hub Poland, Huge Thing, Gamma Rebels, Space3ac, Warsaw startup

Hub accelerator, upstart accelerator, starter rocket, ReacktorX, and Mission to Run are the top known accelerators. ReacktorX helps startups to flourish and aims at bringing different groups such as coders and product designers to collaborate for innovation. Gdansk Business Incubator, UnityHub, punk internet, Brinc, and AD VENTURES are some of the renowned startup incubators present in Poland. The funds for startups were directed to create something innovative and provide distinct infrastructure. Poland has VC ventures like Docplanner, Booksy, Brainly, Cosmose, Perfect Gym, Callpage, Tylko, SethoMe, FinAi, and Neptune that are supporting Poland startups to flourish globally.

Warsaw University of Technology, Wroclaw University of Technology, AGH University of Science & Technology, Poznan University of Technology, Gdansk University of Technology are some of the leading technical institutes that provide infrastructure, accelerator and incubator programs for students and encourage their startup potential. The government of Poland is working hard to bring digitalization in the country. Projects like "From paper to digital Poland" emphasize strengthening digitalization. Tax Relief—The Polish Chamber of Commerce (KIG) maintains a catalog of services for enterprises named KIGNET. It is a support network from regional chambers of commerce to help the competitiveness of Polish enterprises and provides services for small and medium-sized companies.

European Aid grants are direct financial contributions from the EU budget/European Development Fund. Poland's startup ecosystem can create exciting ideas and establish a strong innovative company. Factors like connecting regional hubs, opening them to external collaboration, and encouraging them to innovate are some of the points for Poland to deep focus on.

Russia

"If you wait for the right time or the good times to start a business, you wait all your life."
— **Fran Tarkenton**

Russia, a nation comprising a vast expanse of Eastern Europe and North Asia, is a land of superlatives, by far, the second-largest country in the world. It has an area of 10,672,000 sq. Miles and nearly 150 million inhabitants. Russia shows both monotony and

variety due to its magnitude. There are forty UNESCO biosphere reserves in the nation. The World Bank listed Russia 11th in the overall gross domestic product, with USD 1.5 trillion in GDP in its 2018 study. In Tholons Services Globalization Index 2019, Russia is placed at 7th amongst the Top 50 Digital Nations, having four of its cities included in the index namely Moscow, Nizhny Novgorod,

St. Petersburg and Novosibirsk. St. Petersburg has the best rank of 32 amongst the Top 100 Super Cities. Inc. Magazine voted Moscow the second best-performing city in the world for fast-growing private enterprises in 2018.

The startup ecosystem which has been slowly building over the past ten years has seen noticeable increase in activity. There are more than 3,000 startups active in Russia according to AngelList. Oh My Stats!–an online e-commerce market that analyze businesses and manage marketing budgets; Mail.ru Group–a Russian communication portal which is the largest Internet company in the Russian-speaking world; RocketBank–a mobile app and internet banking that helps consumer to get full control of their finances with high security and support; VitaPortal–an online personalized health content solutions; HipWay–a leading Russian online travel company; Easy ten–a mobile App for learning foreign words, that allows user to learn 7 different languages with 10 words each day; TagBrand, Zet Universe, Voximplant are some of the startups. The startups that attended the "unicorn" state are Avito.ru–a Russian advertisements website; QIWI–an online payment services provider; Ulmart–an Internet-based retailer for electronics; VKontakte–Russia's most popular social network and Yandex–a search engine and service provider.

Moscow is ranked 10th place with the best ecosystems for startups in StartupBlink 2019 report, published by the international information and analytical portal. There are various incubators— Higher School of Economics Business Incubator, StroginoTechnoPark, Branch Agricultural Business Incubator, etc. Moscow State Institute of International Relations is also running a free incubator for students and graduates. In addition to government-sponsored accelerators like the Skolkovo Foundation, a large number of private accelerators and incubator programs are available in Russia. These include programs from firms such as IKEA, French organization MEF, Rutech, NUMA, IIDF accelerator, GenerationS, Disruptive, and global accelerator. In collaboration with an early-stage venture fund and seed accelerator 500 Startups, Sberbank, Russia's biggest bank and a major global financial institution, provides a platform for Russian startups to create their Silicon Valley start. Accelerator REU them. G.V. Plekhanov, the export accelerator of the NTI, Venture Accelerator, 500 Startups, Business accelerator AKSELERATOR.ru, Fashion Futurum Accelerator, Global Growth Challenge, Innovation DeepTech Acceleration Program, Artificial Intelligence Track, EDTECH, Insurtech 2.0 are the top accelerator program provider in Russia.

There are many co-working spaces in and around Moscow-#tceh, StartHub, Co-working 2.0, Work Station, ArmaCoworking are few. There are many grants available such as Innovation Assistance Fund, Skolkovo Fund Granting Program, Venture Investment Development Fund, Innovative Entrepreneurship Support Fund, etc. Many private venture capital firms are also now active, namely Altair Fund, Starta Capital, The Untitled Ventures, Bright Capital, AddVenture Fund, and Internet Initiatives Development Fund.

The government has helped by providing grants and tax rebates on newly flourishing startups. The most prominent government project is the Skolkovo project which started in 2006 and its aim was to create a completely new 'Innovation City' just outside Moscow, complete with top-level university, accelerator labs and all the housing and infrastructure to support a large population. The Skolkovo site also houses corporate innovation groups from Russian organizations like Sberbank and international groups like Boeing.

Part of the government approach has also been to establish privately-run organizations to support startups. Skolkovo Foundation, at the core of the Skolkovo venture, is one of these. Other groups incorporate FII which provides funding and educational support to startups. Also prominent are the Technoparks, set up by city governments, including the 12 business incubators established by Moscow city government containing 1,300 projects across various industrial sectors.

While Moscow and the Skolkovo project are the most prominent hubs of activity, centers of innovation are spread across the country. It has been recognized as a leader in the scientific field. Russians have to work on market understanding so that brilliant ideas can be successfully turned into a global business. This will allow the number of startups to increase and flourish.

Silicon Valley

"It's not magic. It's talent and sweat."
— **Silicon Valley**

Silicon Valley is a tag for the southern portion of the San Francisco Bay Area. The word "silicon" originally referred to a large number of silicon chip innovators and manufacturers in the region. Silicon Valley has attracted entrepreneurs and executives from all over the globe. It has been like the Mecca for talented entrepreneurs. The

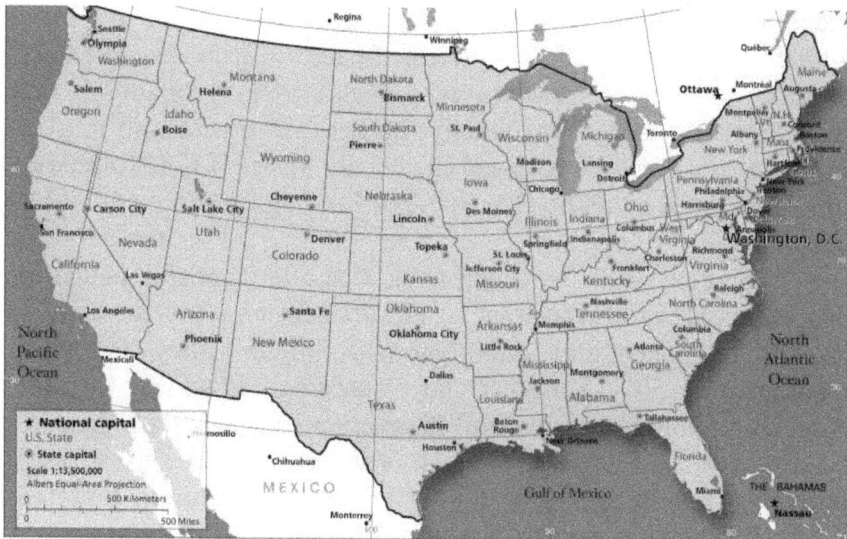

thing that stands out most is the audacity and the grit of the entrepreneurs to go against the market and create products and platforms that people want.

Silicon Valley is headquarters to 39 businesses under Fortune 1000 companies. The area has around 38,500 startups according to AngelList, which include some of the world's largest hi-tech companies. Designlab–an online education platform for students to learn design skills with hands-on projects, live 1-on-1 mentorship from top designers through community interaction; Periscope–a cloud data analysis device that utilizes pre-emptive in-memory caching and statistical screening to perform data analysis; Onfleet–technology company that are specialized in logistics management software and route optimization for businesses; NanoNets–a platform to help developers build Machine Learning models; Golden–a machine intelligence platform that map and generate human knowledge to accelerate discovery and education, Medisas, Distributed Systems, Islands are some of the startups in Silicon Valley.

Silicon Valley startups have the largest representation at the Fortune's "unicorn startups" table. "Unicorns", in the startup world is a startup company valued at over USD 1 billion. Uber is leading the race with a valuation of USD 91 billion, with Airbnb in second place at USD 31 billion. Other popular companies out of Silicon Valley include Snapchat, Pinterest, SpaceX, Dropbox, Theranos, and Lyft. Among all these unicorns, Theranos has probably had the most enduring journey so far. For bigger startups, raising funds is not an issue. New York-based healthcare startup Oscar, already a unicorn, closed a Fidelity-led round of funding, valuing it at USD 9 billion.

Silicon Valley has more than 70 accelerators and incubators programs. Many accelerators in the area host events for startups globally to connect and collaborate with Silicon Valley startups. Y Combinator, 500 Startups, i/o Ventures, Founder Institute, AngelPad, Dogpatch Labs, and Innovation Endeavors are the most successful accelerators in Silicon Valley. Matter, Upwest Labs, Parisoma, Sandbox Suites, Tech Liminal, Sudo Room, Founders Space, kicklabs are some of the incubators present in Silicon Valley. One-third of the country's venture capital goes to Silicon Valley. This made it a leading hub for innovation and startups. Silicon Valley has long established itself as a focal point for investors. SV Angel, 500 Startups, Andreessen Horowitz, DFJ, and Lowercase Capital are a few of the top early-stage startup investors.

Breaking down by industry, Silicon Valley software and life sciences companies both logged the most down rounds. As a percentage of all financing rounds within their categories, life sciences and hardware had the highest representation of down rounds, each at 18%.[79] Investors like First Round's Fred Wilson have put pressure on heavily valued startups, calling for companies like Uber to go public. Public tech companies like LinkedIn chopped its valuation. Marc Andreessen, an investor, foresees an exploding tech bubble. Reports by Fidelity question the valuation of startups. So, the bubble is now gaining everyone's attention.

The US government played an important role to uplift the innovation ecosystem in the region. To encourage small domestic business, the Small Business Innovation Research (SBIR) was setup. They do this by strong leadership, distinguished yet achievable goals, and by giving employees almost free reign on design and innovation. Companies in Silicon Valley have realized that employees come first and that these people make the most difference in the long run. Employees generally expect competitive compensation. However, their aim should be to shape their idea, expand the business and try to become the leader. With this attitude, startups at Silicon Valley have been able to develop products and services that are different and naturally appealing to consumers.

Silicon Valley is a destination on its own. It has evolved as the preferred location to set up technology companies. The place keeps attracting new startups as well. In Silicon Valley, even global tech startups are hoping to get a footprint. Starting a tech company with an existing company infrastructure, skilled resource pool, and a thriving marketplace in such an environment offers a definite head start relative to other places. Silicon Valley's environment is defined by technology, cooperation, and risk-taking. It offers the vital motivational structure that tech startups need.

Singapore

"Technology gives the quietest student a voice."
— **Jerry Blumengarten**

Singapore is an island country off the southern tip of the Malay Peninsula in Southeast Asia. It is a global hub for education, entertainment, finance, innovation and trade. The city is known for its investment potential and has been recognized as the

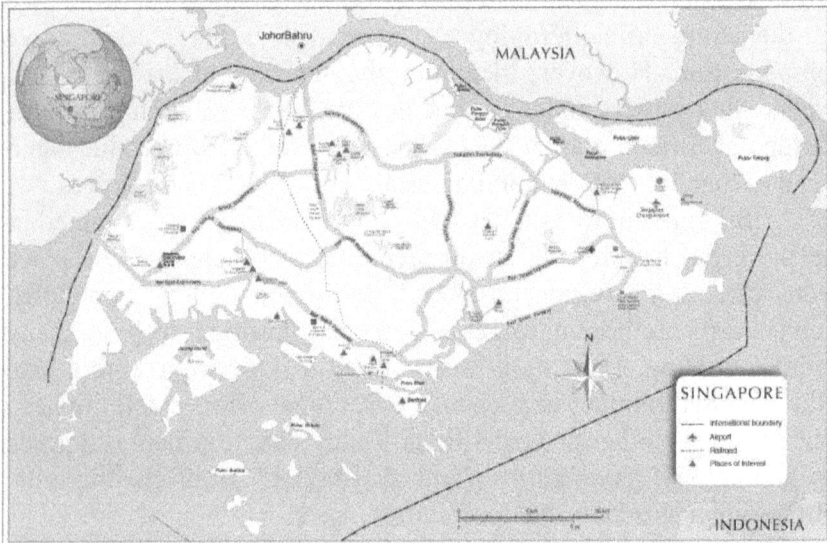

most "technology-ready" nation by the World Economic Forum.

Singapore is known for its advanced and diverse business economy. The main contributors are financial services, manufacturing and oil refining. Export of Petroleum, integrated circuits and computers constitutes 27% of the country's GDP. The 2019 Index of Economic Freedom ranks Singapore as the second freest economy in the world and ranks second in the World Banks Ease of doing business index 2019.

Singapore has become synonymous with 'clean' business and is also renowned for its quality of life. The city is a creativity hotspot and emphasizes work-life balance. Singapore is ranked 25th on 'Mercer Quality of Living Survey'. For young and creative entrepreneurs, it offers a healthy innovation-encouraging environment. Tholons Services Globalization Index (TSGI) 2019 ranks Singapore 9th amongst the Top 50 Digital Nations and 9th place among the Top 100 Super Cities.

Singapore is arguably one of the most tech-savvy nations, not just within Asia, but across the world. Despite being an island nation inhabiting just 5.6 million people, Singapore is one of the world's leading centers for technological innovation. With its luring tax breaks and developing market opportunities, it is a go-to place for multinationals.

The nation's startups ecosystem has made significant progress over the past decade. Singapore is ranked 12th among the top startup ecosystem globally. A report published by the US-based Startup Genome project, which covered 10,000 startups and 300 partner companies, stated that Singapore has overtaken tech mecca Silicon Valley as the world's number one for startup talent.[80] The country's innovative policies, significant government subsidies, ease of starting businesses, and geographical location paved way for its success.

With over 6,000 tech startups, 100 incubators, accelerators, venture builders, and over 150 VCs, the city has a robust base for startups. Lazada–e-commerce platform, Sea–Fintech, Air Trunk-Telecom, Trax–Image Recognition, Bigo–Broadcasting are some of the top startups. The city has two unicorns—Grab, a *company* offering ride-hailing transport services, food delivery, and payment solutions; Sea Limited, a consumer internet company.

The future of startups in Singapore is predominant. Startupbootcamp, Citi, and Wells Fargo are amongst the 50+ organizations running accelerator programs in Singapore. Sector Specific Accelerator (SSA) Programme–aids startups in the medical and clean technology sectors. The program commits up to SD (Singapore Dollar) 70 million (USD 51.34 million) annually to support these startups. Singapore's government pumped SD 19 billion (USD 13.94 billion) into technology with the potential to make Singapore a lucrative location for entrepreneurship and innovation. "Startup SG" is an umbrella brand by the government that unifies all schemes and incentives to uplift startups culture. ACE

Startup Grant—supports first-time entrepreneurs with mentorships and startup capital grant.

In Singapore, startups enjoy tax exemptions for the first three years whereby they pay no tax on the first SD 100,000 of the chargeable income and only 50% of the tax on the next SD 200,000. Apart from the low tax rates, the maximum tax being 17%, and the ease of doing business. Another important factor that attracts entrepreneurs from around the world to the country is the multiplicity of financing sources for startups.

The presence of multiple Venture Capital Firms who can support late-stage businesses has been one of the unparalleled features of Singapore. Singtel Innov8, Golden Gate Ventures, IMJ Investment Partners, Wavemaker Labs, Ardent Capital, Jungle Ventures, Sequoia Capital, 500 Startups, and Life.SREDA are some of the major ones. Universities in Singapore have been organizing startups pitch competitions for a long time. Nanyang Technological University's Ideasinc, the National University of Singapore's Global StartUp are the leading pitch events and incubators that are being run.

Singapore has remained a center of attraction and a leader in the region for finance and digital. Asia has a huge untapped market and Singapore provides the best infrastructure, talent, and environment for entrepreneurs to capture this market. This is where Asia's entrepreneurs and top talents are looking to realize their ideas.

Sweden

"If GM kept up with technology like the computer industry has, we would all be driving USD 25 cars that got 1,000 MPG"
— **Bill Gates**

Stockholm is becoming one of the universal IT hubs in Sweden which has almost as many new businesses as Silicon Valley does. With a population of nearly 10 million, Sweden is in its league to establish innovative technology. This Scandinavian country is home to some of the world's most popular brands, including IKEA, Volvo, Electrolux, Ericsson, and H&M. Sweden ranks 7th among 202 nations globally, based on the strength of its startup ecosystem according to StartupBlink 2018 report.

The nation with ten million people is making tech tick. Worldwide successful startups like Candy Crush, Minecraft creator Mojang, and Spotify all originate from Stockholm. We know about huge organizations like Saab, IKEA, H&M, Huskvarma, Volvo, and Ericsson, but we wonder how Swedish new businesses like Linux, Skype, MySQL, Wrapp, Memeto, and Klarna get so big. Nordin tech generated revenue of USD 2.9 billion per year from exports from which a greater part arrives from Sweden.

Around 1,566 startups are based in Sweden according to AngelList. Bloglovin'–an online fashion and lifestyle app of Sweden; Mapillary–a street graphics platform to scale and automate mapping that generates information for map improvisation, city development, and automotive industry advancement; Campanja–a technology-driven company that uses advanced technologies such as real-time tracking, predictive modelling and high frequency bidding for online advertisement; Lifesum–the most popular digital health service of Nordics; Fishbrain, Brisk.io and Anyfi Networks are some of the renowned startups in Sweden. The major beneficial project backed by the capital is "The Spotify" in Europe. Spotify motivates a modern generation of business people and empowers blessed financing from overseas; Payments processing company "iZettle" acquired by PayPal; Klarna and Northvolt are the only companies in a long line of a successful startup with the name "Unicorn." "The music streaming service" is the name provided to startups that were established after 2003 that achieved a value of USD 1 billion and is termed as the fifth unicorn. According to the Atomico report, the investment firm, Sweden is the second nation (based on per capita) that produces 6.3 billion-dollar companies per million people, compared to Silicon Valley's 8.1.

Sweden consists of multiple accelerators/incubators. 500 startups, THINK Accelerate, IKEA Bootcamp, SICS Startup Accelerator, Fast Track Malmö Accelerator, Lund Sales Program and more. The accelerator searching for new businesses across the globe to take part in its startup accelerator is the Fast Track Malmo. It is a collaborated project by ALMI Invest, Malmo startups and invests and Minc in Skane. The startups selected will be assisted with R450,000 to R750,000 (Rands-South African Currency, approx. USD 30,000 to USD 50,000) along with free space for office and access to 40 companies and 300 people. They will also have access to investors, angels, mentors, and a global network of resources. Swedish Incubators and Science Parks (SISP), Wallstreet Ventures, Amplify, KTH Innovation, Sting, SU Incubator, Arctic Business Incubator, Umeå Biotech Incubator, Inkubera, Young Capital, SUP46 are some of the renowned incubators.

According to a report published by the Stockholm-based investment firm Creandum, the Nordic region represents two percent of Global GDP but has accounted for almost ten percent of the world's billion-dollar exits over the decade, with more than half from Sweden. In terms of exit value, the 2014 year was noted as the best year for Nordic companies with USD 1 billion exits and more than USD 13 billion in total exit value.[81]

This rich history of global brands laid the foundations for the Swedish government to invest heavily in its technology infrastructure in the 1990s. Over 60% of the nation has access to high-speed fiber optic internet with rates of 100 megabits per second. Swedes are preparing with the advanced savvy and physical tools in becoming the nation of ready consumers and country of disruptors. Swedes also cherished the era of individuals who grew up on the internet, developing a culture of open access and entrepreneurial collaboration.

On 1 December 2013, a special tax incentive scheme encouraging investments in small businesses was applied. Individuals acquiring shares in small companies at the formation of the company or through a new issue of shares can deduct 50% of the payment against capital income.

Niklas Adalberth and his company Norrsken Foundation are living proof of how Sweden is currently functioning as a tech hub. The main aim of the non-profitable foundation is to guide the methods and tools of entrepreneurship, like product development, big data, and apps, into social impact areas like charity and corruption. The startup's support uses two apps that guide the migrant to find new jobs, called "Just Arrived" and "Welcome App." Another is Klarity, founded by Adalbert which is used for reporting corruption and enables the user to upload images of it.

Stockholm Fintech hub is the first Fintech hub launched by Sweden. Stockholm's startup scene is ready to receive a thrust in the upcoming years. Stockhold Fintech hub will provide a support system that can minimize challenges and increase opportunities for entrepreneurs in the area. *"We know there's going to be challenged with the Fintech companies, the banks, and regulators both here in Stockholm and Brussels. But if we strive to create an environment where they are lessened, we will be a lot further ahead than we are today."*—Stockholm Fintech Hub co-founder and BLC Advisors partner Matthew Argent. The aim of the modern hub is for professional knowledge to be shared and teamwork to be created between large banks, entrepreneurs, regulators, investors, and the Swedish government. It will be shared with 42 Fintech hubs across the globe and partner with identical Scandinavian hubs in the vision to help the developing startups.

The central Stockholm establishment is having a conference center, a co-working space and shall provide education programs for financial technology startups. For intervening banks to provide Fintech entrepreneurs with what they are searching for in terms of innovation is the plan of such an information-sharing event. In addition to the Fintech hub, a 170 million kronor hub is to be launched supported by the Klaran to target entrepreneurs who are likely to find solutions to social problems.

The nation has a "all for one and one for one" ethos that operates without bureaucracy and as an even unit. The business is as great as its individuals and this is realized by the Scandinavian companies which make Sweden one of the best technology hubs out there.

Switzerland

"What new technology does is create new opportunities to do a job that customers want done."
— **Tim O'Reilly**

Switzerland is recognized for its tourist attractions and picturesque scenes worldwide. Its natural geography not only has mountains but also includes central highland of rolling hills, plains, and large lakes. In comparison to other European countries, Switzerland is more artistically diversified. It has four official languages: German,

French, Italian and Romansh. Switzerland's vital location in the core part of Europe has made it an economic crossroads for hundreds of years. The aristocratic framework, including the Gotthard base tunnel, is 57 kilometers long, serves to promote trade among the largest economies within Europe. This has helped in transforming

Switzerland into one of the continent's biggest supplier hubs. Switzerland is ranked first in the Global Innovation Index 2019 and is named the world's most innovative country.

AngelList lists more than 1,967 startups in Switzerland. eWise–a leading international banking app that provides privacy and security for users data; Swiss Founders Fund–a seed and early-stage investment fund; Coteries–agency that design, develop and market highly qualitative digital products; Designique-Swiss interior design shop; Dolphin Engineering–PreDiVine (model) that predict crop diseases, using sensor networks and bio-models and suggests treatments; iceVault–an offline storage app that secures online digital data like bank details. The rambling epicures, ShapeShift are some of the renowned startups. Switzerland has 4 unicorns on its list. MindMaze is the first Switzerland-based unicorn. It serves as the platform in developing instinctive man-machine interfaces with the blend of VR, brain imaging, computer graphics, and neuroscience. Roivant Sciences–A pharmaceutical company that builds subsidiary biotech and health technology; Numbrs–A Financial Technology Company that aggregates the customers bank details ad credit card information; Avaloq Group–Software developer and provider for core banking.

Zurich, Vaud, and Bern are the main startup hubs in Switzerland. The success of the Zurich and Vaud startup scene is benefited from the presence of globally renowned universities Swiss Federal Institute of Technology Zurich (ETH) and EPFL. Zurich and Vaud have always had the maximum number of exits in Switzerland. One of the most habitable cities in the Mercer report, Zurich is ranked among the top three places and is home to the largest engineering office. It has a rich accelerator and incubation support system for early-stage startups. The IBM Research campus is home to two Nobel prizes. ImpactHub Zurich, Startup Grind Zurich, Zurich Entrepreneur meet, Hack Zurich, Investor Summit, and World Web Forum are the major startup events organized in Zurich. Kickstart Accelerator, CTI Startup, STARTUPS.CH Academy, and Swiss Startup Factory are some of the successful accelerators/incubators. Large companies of the country, including media, transportation, financial, retail and many others, supported the initiative known as Digital Zurich 2025. To encourage the progress of Zurich as the place to initiate digital ventures is the main goal of the initiative.

The provinces that show the desire to offer incentives for startups in their region are backed by the government which provides a guarantee for federal loans. The remunerations on grants have been redistributed and the provinces have the freedom to grant financial and other types of incentives to startups. Domestic and foreign investors have equal eligibility to benefit from remunerations. New companies are eligible to receive tax breaks if the investment on their assets is 50%. Moreover, some provinces have restrictions on tax breaks as it is against federal tax liability. The maximum tax break is 50%. The startups and entrepreneurs obtaining bank loans are backed by the Swiss government by offering commercial guarantees. Loan guarantees are available for up to 500,000 Confoederatio Helvetica Franc (CHF) and the Swiss Federal Government underwrites 65% of the debt exposure. [82]

The startup environment is supported by admirable education structure, funding, top research institutions, and an overall combination of industry, academia, and politics that work to bring innovations to the market. The Swiss dual-education system is unique worldwide. It unifies internships in an organization and professional education in schools. With its widely recognized universities, Switzerland is one of the most competitive countries in the world with a pool of highly educated graduates with deep know-how in the fields of finance and technology. Switzerland has 60 higher education institutions, of which four universities are among the top 100 of The Times Higher Education World University Ranking, with the Swiss Federal Institute of Technology Zurich (ETH) ranking #7 in 2019.

Switzerland's innovative startup scene is thriving, with some of the world's renowned educational institutions and the R&D centers of Google, Apple, Facebook and Amazon being present in the country. Switzerland provides an optimal ecosystem for technology startups, but the startup scene has not yet succeeded in attracting the interest of foreign investors.

Texas

"Never before in history has innovation offered promise of so much to so many in so short a time."
— **Bill Gates**

Texas has proved to be a good place to do business due to its comparatively flexible regulatory ecosystem, without sole or associated income tax and prosperous growth in several cities. Texas stands as the second-largest state with USD 1.6 trillion gross domestic products (GDP) after California. The small businesses in

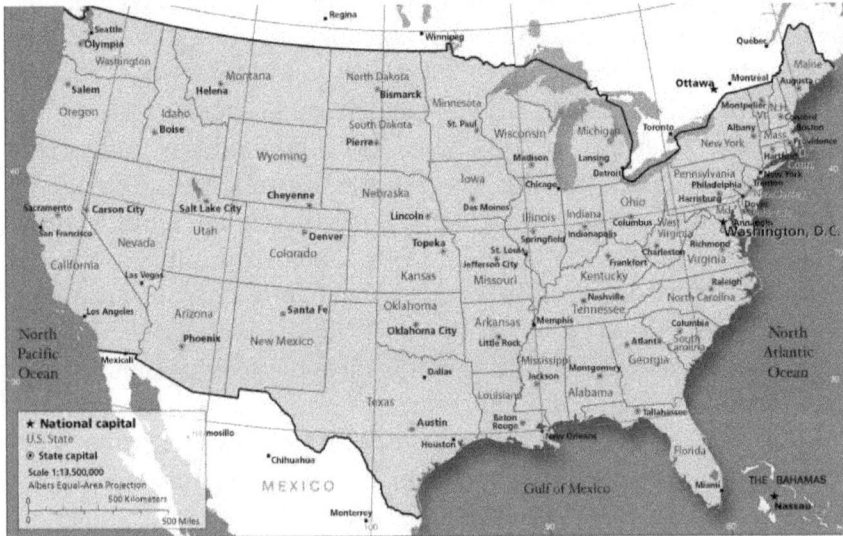

Texas have employed over 45% of the private workforce in 2018. In Tholons Services Globalization Index 2019, San Antonio, Texas ranks 34 amongst the Top 100 Super Cities. According to the report by Forbes, Dallas is among the top five upcoming cities to be the "Silicon Valley Tech Hub." The enormous growth in entrepreneurial and technical activities is being witnessed in the region of Dallas. Easy access to reasonable real estate is one of the major advantages of the region. According to US News 2019, Texas is ranked #1 Best Place to Live in the USA. Austin Ranked 5th Best City to Start a Career according to WalletHub's report, May 2019.

The flocking of businesses, real estate developers, and families to the region is increased as there is an increase in opportunities for job and innovation. Texas has over 12,333 startups in AngelList. Some of the renowned startups in Texas are: Proof–a web based online marketing solutions; Living Security–a cybersecurity coaching firm teaching employees on how to reduce cyber risk; Pushnami–a leading digital marketing software provider in Austin; Shipwell–a tech startup that connects the fragmented of freight industry and helps customers instantly book and track freight shipments under single roof; ALTR–a secure blockchain-based data storage for commercial use; Better Living Technologies–a public company that assists individuals with type 1–diabetes; Backtracks, Blink Identity, Care2Rock, Bungalo, WP Engine are few startups in Texas.

Texas has over 60 accelerators and incubators in its list. Capital Factory, The Business Factory, NTEC, TECH Fort Worth, iStart Valley, Launch, Small Business Development Centre (SBDC), Entrepreneurship and Commercialization Centre's (ECC), VelocityTX, WT Enterprise Centre, EGBI are some of the renowned startups accelerators in Texas. The incubators in Texas are IdeaGist, Services Cooperative Association, Coastal Bend Business Innovation Centre, The Hub of Human Innovation, Austin Technology Incubator, LAUNCH, DFW, Fort Work Academy, Epoch Theory, SATUS, NoD Coworking, and Velocis. One of the innovations of Texas Tech, Kinetic Accelerator is moving into the Research Park. It supports the setting up of new businesses and explores licensing opportunities based on inventive technologies developed from research labs NASA, Texas Tech, and many more.

There is an increase in modern developments and in creating room to extend their businesses for tech companies in the region. Dallas' new businesses venture capital raised USD 639 million in 2013 which was more than the VC of Huston and Austin combined. Tech Wildcatters, Health Wildcatters, 2M Companies LLC, Ward.Ventures, Aristos Ventures, Mark Cuban Companies, Maverick Capital are some of the renowned Venture capitals in Texas.

University of Texas, Austin is ranked 49 among top public schools nationwide as per US News and World Report, 2019. Apart from cities like Dallas, universities are also leaping into the technology hub world. Texas Tech University Research Park officially opened in August 2015. The conventional laboratories and computer visualization are offered under Leadership in Energy and Environmental Design (LEED) certified facility. The advancement towards becoming the major innovative national research university, Innovation Hub and Research Park serve as a major component in Texas Tech. It also serves as a key for the academic and economic platform in West Texas. Groups, businesses/individuals can loan office space in the facility to carry on research, making it unique from the rest of the Texas Tech campus. Texas Tech Innovation, Mentorship and Entrepreneurialism (TTIME) organization, created under Texas Tech, will also have its base at the Research Park. TTIME is dedicated to supporting students in advancing ideas of entrepreneurship.

Another city in Texas is being looked upon for innovative progress in Austin. The US Department of Defense has selected Austin as the third location of its in-house tech startup, the Defense Innovation Unit Experimental/DIUx. The Defense Department created DIUx in 2015 to develop deeper connections between the military and the US technology industry. DIUx is split into three teams-Its "Venture Team" identifies and explores the battlefield impact of emerging commercial technologies. Its "Foundry Team" is focused on technologies not yet fully developed for the battlefield, but could potentially have military applications. Its "Engagement Team" works to connect military leaders with entrepreneurs. Unlike venture capital firms that seek equity in growing companies, the DoD ill only be purchasing intellectual property and prototypes. Houston is not far behind in the tech hub either. BrainCheck can identify a possible impact based on games played on a tablet within a short time. This company, guiding its products in charter schools of "The Knowledge is Power Program" (KIPP), is advancing its exposure and is monitoring through a business accelerator at Texas Medical Center, where BrainCheck is working together with other concussion researchers to improve the science around head injuries. BrainCheck is one of the companies in the leading class of "The Texas Medical Center" (TMCx) among the other 22 companies.

Announced in October 2014, TMCx is part of the Texas Medical Center's push to become one of the world's premier life science commercialization clusters. For the companies beyond the accelerator phase, TMXx+ was started and will complete the framework of Johnson & Johnson Labs JLABS, which is the incubator by Johnson & Johnson. JLABS works by providing life science entrepreneurs with shared lab space, private office, and modular laboratory suites, as well as state-of-the-art equipment and value-added operational, education and business services, all vital lifelines for startups struggling to gain enough funding to get off the ground. The Houston site, being the pioneer in medical device prototype lab like the 3-D printer, provides access to well-qualified tools, and skills-developing programs to develop and design advances in health technologies.[83]

The enterprise in the space of Houston extends up to the Office of Technology Commercialization (OTC) and the University of Texas MD Anderson Cancer Center, supported the setting up of biotech startups. Since 1987, OTC has been involved in the creation of 11 affiliated life science companies that have raised more than USD300 million on the strength of MD Anderson technologies, including ApoCell, Castle Biosciences, and DNATrix. Four portfolio companies listed in Nasdaq have raised more than USD 230 million and funded USD 25 million in sponsored research at MD Anderson.[84] There is no doubt that Texan cities are becoming tech hubs on their own merits. As Silicon Valley gets saturated, VCs are looking into other options to invest and Texas is proving to be a good contender for investments.

UAE

"Technology can become the "wings" that will allow the educational world to fly farther and faster than ever before—if we will allow it."
— **Jenny AR ledge**

United Arab Emirates (UAE) is a federation of seven states and is the second-largest economic region of the Middle East. UAE has the seventh-largest oil reserve in the world, and one-third of the GDP of the nation comes from the oil industry. The main cities and seven emirates are connected by multi-lane highways, with several

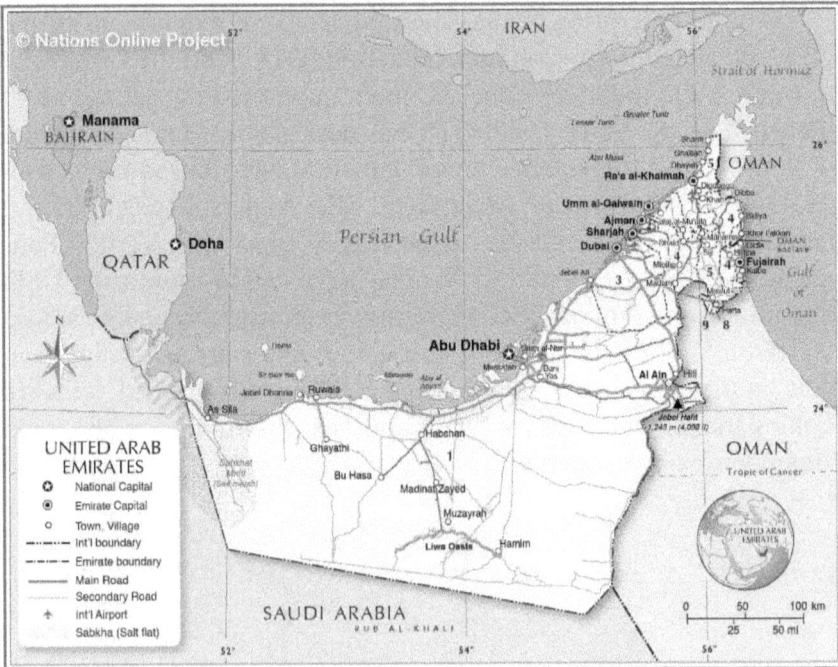

international seaports and airports. This serves millions of travelers and thousands of ships and cruises every year. HSBC's Expat Explorer 2019 survey highlighted Dubai as the 9th best city globally for expat entrepreneurs.

Tholons Services Globalization Index 2019 ranked Dubai 26th among the Top 100 'Super Cities'. 2,188 UAE startups are listed on AngelList. MidChains–an institutional crypto asset trading platform that provides fully regulated and supervised ecosystem infrastructures for digital asset like trading and investments; Cartlow–a mobile App that provide simple way to buy Pre-owned mobile with certified products, warranty, and easy returns; ZIQQI–an integrated b2b e-Commerce wholesaler platform; One Moto–an electric motorcycle brand, that are designed and developed for inner-city traveler in UAE. An app-based car service, Careem Networks, is a unicorn startup from UAE that connects the public consumer to available rides in minutes, at the tap of a button.

Several of these startups have become huge regional success stories; one of such startups is e-commerce giant Souq.com. Souq allows thousands of buyers and sellers to buy and sell a huge variety of products over a trusted platform. Souq was successfully acquired by e-commerce mega-giant Amazon, for USD 580 million in March of 2018. With over 3,000 employees, the company is the region's biggest tech success story. The UAE has emerged to be a preferred destination for innovators, talent, and entrepreneurs from across the world, stated by professional network LinkedIn in the study.

Startups in the UAE are fostered by several accelerator and incubator programs established by both public and private contributions. Turn8 and Flat6labs are two major accelerators in UAE. The Turn8 program tactically hunts for new ideas and thoughts that can be cultured and introduced to the market through the Turn8 seed accelerator. The program commences with an accurate online and event-based international inspection campaign, tailed by a 120-day financed "Rapid Fire" Business Modeling and Prototyping seed accelerator, and lately an Investor Demo Day.

Abu Dhabi and Dubai are two major startup hubs in the UAE. The startup founders in Dubai, from Careem to InternsME, an online video resume career portal for internships and entry-level jobs, are growing every year. Dtec, based in Dubai, is a global hub for innovation and entrepreneurship. Being a thriving place for over 500 startups and SMEs and situated in the central part of Dubai's Silicon Oasis, it provides easy-going work solutions for innovative entrepreneurs. Dubai is putting itself at the forefront of business, science and public spending, with an accelerator program that aims to draw 30 of the best startups, from around the world to the city to tackle the most important public problems. Dubai is awarding government contracts of approximately USD 30 million to startups.[85]

Abu Dhabi is home to more than 100 tech startups. With accelerators and hubs such as twofour54 and Flat6Labs, the city provides promising infrastructure for tech startup growth. New York University, Abu Dhabi, has launched many programs to support the regional innovation ecosystem. Their Digital and Business Literacy program provides practical technology and business skills that are essential for aspiring entrepreneurs. A ten-day program offered by the university focuses on helping Fintech Startups with training and tools to evolve a complete modern project that is investment-efficient, adaptable, and scalable. The financial technology hackathon hosted by GlassQube in partnership with Abu Dhabi Global Markets and Temenos on December 8th,2016 turned out to be a huge success. Over 100 people participated across 10 teams and the caliber of the participants and their output was world-class. The event officially ranked as one of the top GFE hackathons ever held anywhere in the world. The startup culture in Sharjah is also seeing continuous government support. The startup-centered initiatives like government-sponsored Sheraa (The first university-based accelerator of Sharjah), the Sharjah Chamber of Commerce's Shjseen, and the AUS (American University of Sharjah) Research, Innovation and Technology (RTI) Park, showcase the engaging approach of government to attract and retain tech talent.

In recent years, the startup scene in Saudi Arabia and Jordan has been growing at a steady pace. To decrease its financial dependence on oil, the Kingdom of Saudi Arabia declared many economic diversification initiatives in 2017. According to the economic plan "Kingdom Vision 2030", the government of Saudi Arabia aims to take the oil titan Saudi Aramco public, and move the business from an oil-producing company to a global industrial conglomerate. In 2017, Japan's SoftBank Group and, Saudi Arabia's sovereign-wealth fund, launched the world's largest technology fund, a nearly USD 100 billion vehicle that steers capital to cutting-edge technologies in the US startups and other global firms.[86]

Jordan is attracting the investors' interest by its innovative tech-based startups. The REACH Program was Jordan's first national IT plan, developed with input from various investors, supervised by King Abdullah, and executed completely by the private sector. The Queen Rania Center for Entrepreneurship, a non-profit organization is also playing a major role in fostering startups in Jordan. Oasis 500 is another very successful accelerator program in Jordan which was established in 2010. Since then Oasis 500 has fostered some of the well-known Jordanian companies like ArabiaWeather, e-payment platform MadfooatCom, and bookseller Jamalon.

Robust government support is available to SME's in UAE. To motivate local entrepreneurship, the government's Khalifa fund of around Dh 500 million was established. It provides interest-free loans for seven years, counseling, training, and after-funding support. The UBI Index positioned AUC Venture Lab as one of the Best Performing University Business Incubator in North Africa and the Middle East.

Over 30 Free Trade Zone (FTZ) is present in the United Arab Emirates. The enterprises are liable for guaranteed tax holidays for a period of 15 to 50 years and grant dispensation from tariffs and excise. The UAEU Science & Innovation Park Business Incubator supports the UAE vision and the UAEU SIP vision and mission to move towards an innovation-based economy. Highly integrated into the UAE entrepreneurship institutional ecosystem, and leaned against a network of international incubators, the UAEU SIP Business Incubator offers its fellows the required infrastructure, coaching and mentoring services to support them in launching a startup. With a focused and well-strategized approach by the public and private sector to promote entrepreneurship, UAE is well set to become the leader.

Uruguay

"Great things in business are never done by one person. They're done by a team of people."
— **Steve Jobs**

Uruguay, with a population of over 3.4 million, is one of the smallest countries in the world covering 175,020 square Kilometer land area and is called the "Switzerland of Latin America." Montevideo, the

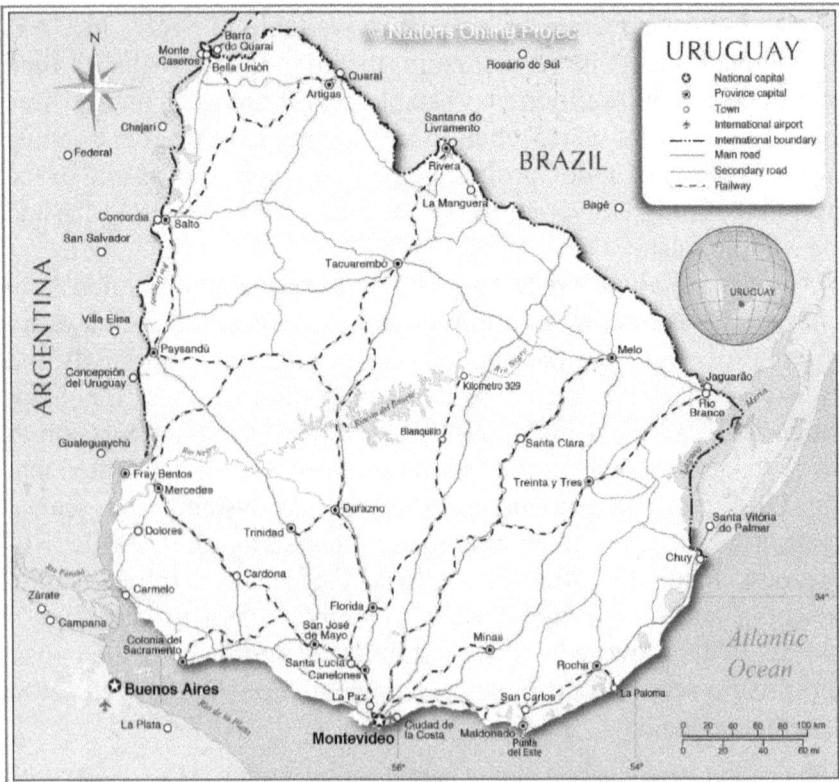

capital city, constitutes highest population of 1.3 million. The quality of life is very high and has a vibrant cultural center in South America. The UN considers it to be a high-income country.

In Latin America, Uruguay stands out as an egalitarian soci ety and because of it's high per capita income. In Tholons Services

Globalization Index 2019, Uruguay ranks 18 amongst the Top 50 Digital Nations, and its capital Montevideo features at rank 14 amongst the Top 100 Super Cities.

With a bit of hard work and ventures, they created apps, programs and services companies that achieved outcomes just as astounding as their athletic accomplishments. Software development is ranked second in the country's exports. Uruguay has begun to attract large names such as Microsoft. It offers the ideal chance for software business, providing tax exemptions to those who export their products. The use of Open Source in the public administration and its laboratory for open data experiences make Uruguay as one of the most advanced country in the world.

Uruguay is a gold mine for foreign investors. Though the market is on the smaller side, the business incubator scene is thriving, and the economy has been steadily growing by 5-6%. Uruguay has 202 startups under AngelList. Fastcall–a high-rated softphone on Salesforce AppExchange; MonkeyLearn–natural language processing platform; Widow Games–board games development firm; Crafted–internet and mobile consultancy store; Xmartlabs–Mobile design and startup engineering; Sophilabs–a software design and development agency, are some of the few blossoming startups in Uruguay.

Accelerators and Company Builders play a significant part in the early phase ecosystem, offering financing and opportunities for entrepreneurs who may not otherwise have access to capital. The 2016 LAVCA Accelerator Directory involves Latin America-based certified accelerators. In 2017, "500 Startups" in partnership with ANII launched a new accelerator program. Through six weeks of mentoring, development courses and relations with shareholders and corporate associates, the "500 Montevideo" Accelerator Program trained 20 Uruguayan and global startups. 85 Labs, Access Latina, Softlanding Uruguay, Socialab are few of the accelerators that are funding startups in Uruguay. In Montevideo, there are a chain of incubators that assists the tech scene. The first technology business incubator in Uruguay was launched by Universidad ORT, one of the largest private universities in Uruguay. The incubators focus on developing entrepreneurs and enterprises in fields of computer science, telecommunications, and biotechnology. DaVinci

Foundation is an incubator that provides 6 months incubation program.

In 2013, Uruguay launched the PAFE program in collaboration with Inter-American Development Bank to assist aspiring entrepreneurs financially. The main idea behind the program is to attract the private investors. There are several incentives for entrepreneurs to set up business in the country. NXTP labs, Prosperitas Capital Partners, Iron Foundry, Tokai, and Kaszek are few among the active venture capital firms in Uruguay. The presence of many co-working spaces such as SinergiaCowork, CoWorkLatam, Serratosa, You Hub, and GepianCowork is supporting the startups growth by providing world-class facilities and allowing businesses to focus on building out their solutions.

Uruguay's to-do-list can mark a tick on most sectors. It's an open, steady, pro-business, and pro-inventive environment. The Uruguayans are testing new technology solutions and are all set to come to the global scene with a bang.

Vietnam

"Really in technology, it's about the people, getting the best people, retaining them, nurturing a creative environment and helping to find a way to innovate."
— **Marissa Mayer**

Vietnam is one of the rapidly growing economies of Asia. Vietnam has been making rapid progress to achieve its vision of becoming a fully industrialized nation. The World Bank has ranked Vietnam at 68th place in its "ease of doing business index" in 2018. Tholons Services Globalization Index 2019 ranks Vietnam as 13th among Top 50 'Digital Nations'. Ho Chi Minh City (39th) and Hanoi (31st) are ranked among the Top 100 'Super Cities'. The rapid growth of the Vietnam economy is majorly contributed by its Foreign Direct Investment (FDI) attractiveness and the private sector. The country has long been an attractive FDI destination, particularly for Japan, South Korea, Taiwan, and Singapore. FDI inflows an average of 8% GDP annually, the highest among major emerging markets in ASEAN. More than half of the total FDI stock is in manufacturing.

Vietnam is one of the fast-blooming tech startup hubs in the ASEAN region. The country is now home to 3,000 startups, making it the third-largest startup ecosystem in Asia. FPT Shop–an online retail platform for electronics; Momo–a mobile payment app that

allows users to pay online and transfer money; Sendo.vn–an e-commerce site of accessories. Tiki.vn–a digital retail platform; Coc Coc–a mobile web browser that provides users with local services and information; Topica, YOLA, Vntrip.vn and Leflair Vietnam, are some of the renowned startups in Vietnam.

VNG Corporation, with over 20 highly popular entertainment, community and software products, is the first unicorn from Vietnam. The CEO, Le Hong Minh, founded VNG in 2003 in Ho Chi Minh. Minh managed to bring home Kingsoft, a Chinese software company with a well-known game Swordsman Online, to license the game to vinagame. In 2009, Vinagame changed the corporate image as VNG and variegated from being an Internet gaming company.

Vietnam has more than 25 business accelerators/incubators. Vietnam Silicon Valley, Founder's Institute, Egg Agency, X-Incubator, IDG Ventures Vietnam, DFJ Vinacapital, Vuon Uom Vat Gia, CyberAgent Ventures, PVNI, Hatch! Program, Hub.It, VIISA and Alphavision Angel accelerator are some of the renowned incubators and accelerators. Vietnam Silicon Valley project, an initiative of the Ministry of Science and Technology and the Vietnamese government, provides startups' numerous programs under the guidance of international and local mentors. As on 2019, Vietnam Silicon Valley has 75+ startup investments. Accelerator "500 Startups" have USD 10 million Vietnam-focused Funds. The USD 10 million target funds point to make 100-150 speculations into Vietnam-connected new businesses. Standard Chartered Private Equity, which has stakes in retail, agriculture and entertainment companies in Vietnam, invested USD 25 million in e-wallet app MoMo. Goldman Sachs also participated in this round with USD 3 million.

The country has the third highest rate for startups in Southeast Asia. The startup community is majorly driven by the country's prime cities Ha Noi, Da Nang, and Ho Chi Minh. Ha Noi produced many tech companies such as FPT, VNG, VC Corp, Vat Gia, and others. Da Nang is the powerhouse for economic growth in the central region. The city is well known for its low-cost environment, strategic location, developed infrastructure, robust engineering talent pool. Ho Chi Minh is the biggest business hub in the country. It is at the forefront of the innovation wave and home for over 50% of the country's startups.

Turner Asia Pacific, Time Warner's Asian broadcasting group division, had purchased a significant stake in Vietnam's digital media and POPS Worldwide creator network. Constituting over 90% of the domestic online music industry, POPs are operating with over 1,700 content associates on more than 1,200 network channels. There were 67 startup investments in 2015, and Vietnam saw a rise of 130% over 2014 and investment further grew by 46% in 2016. According to Topica Founder Institute (TFI), in 2018, the total investment value in Vietnamese startups was up to 889 million USD, three times higher than in 2017.

NATECD (National Agency for Technology, Entrepreneurship, and Commercialization Development), NATIF (National Technology Innovation Fund), IPP (Finland-Vietnam Innovation Partnership Program), mLab East Asia, Hoalach Hi-Tech Service Center are some of the government-aided programs for Vietnam startups. Special tax plans for startups were also endorsed by the Vietnamese government. The taxes imposed on Vietnam are of the national level. Corporate Income Tax (CIT) is levied on local profits of companies operating within the country at a rate of 20 percent. Every year, CIT is payable. The two preferential rates of 10% and 20% are accessible for 15 years and 10 years, respectively, beginning from the start of incentivized activity income generation of young startups.

One of the major advantages that Vietnam provides for technology startup in the country is the availability of highly skilled and low-cost talent pool. The increasing public and private sectors' emphasis on making Vietnam a tech innovation hub and the presence of highly skilled tech labor pool are slowly but robustly shaping Vietnam's innovation ecosystem as a preferable location for innovation and tech-based investments.

23

Conclusion

Destiny of every business is in their own hands. This statement has never been more true. The digital transformation and disruption catalyzed by the myriads of innovation is challenging each business to transform or sink into oblivion. Everyone is keen to embrace the realities of a re-imagined consumer experience, innovation, intelligent automation and the inevitable change. "If the rate of change on the outside exceeds the rate of change on the inside, the end is near."—Jack Welch, CEO, GE.

The most powerful and profitable revolutions require advancements across the entire organization, at all levels—from the individual worker to the C-suite, and across all the associated business components, such as processes, activities, and resources.

We have more mobile devices globally than the count of human beings. Tech savvy business people are successfully capitalizing on this opportunity to market their products and services. Failing to have a mobility strategy can lead businesses to significantly lag behind their competitors. In this age of social media and mobility, businesses can no longer neglect, even the 5% of unsatisfied customers. Based on the social media reach, those 5% customers can degrade the brand name and can hammer the organization revenues significantly.

Cloud technology is taking business transformation to whole new level. "Today's leaders need to be business transformers who leverage the cloud to accelerate and enhance core business processes, through powerful mobile first, software as a service application."—Bob Evans, Chief Communications Officer, Oracle.

In the past 25 years, the digital revolution has taken industries by storm, and no sector is immune to disruption. This technological hurricane is forcing economies to be reshaped overnight as digital disruption brings down flourishing businesses to ground zero in no time. New digital technologies have given birth to a new business age—the "Age of Innovation." Technology is redefining industries and enabling enterprises to diversify their product and service offerings. The only means to survive this competitive business ecosystem is to "re-imagine and re-design" consumer and client experiences.

Innovation at scale for enterprises is a digital transformation process/function which brings three key elements to deliver sustainable impact:

1) Re-image Consumer Experience—Enterprises are working with creative design studios to re-imagine consumer experience through a process of "dreaming." The idea is to take clients and consumers to a state of levitation that can enable them to dream the impossible. It is important to ensure that the dreamers are free from the shackles of technology, practicality and implementation realities. This creates a sweet but near impossible dream. How do we make the impossible, possible!

2) Bring in Innovation—The world is full of innovators, who have re-imagined, re-designed and implemented amazing innovative solutions to deliver snippets of near impossible dreams. It is possible to explore, discover and bring in these innovations to make the impossible dreams, near possible. The solution design through innovation in most cases delivers the wow consumer/client experience. The innovative solution includes data and intelligent automation strategy, cloud, mobility, blockchain, AI, social media, IoT, machine learning and cybersecurity with advanced analytics to unlock new intelligence. The solutioning augments the power of humans with artificial intelligence, leverage AI and advanced analytics algorithms to sense, comprehend, act and learn across value chain at an unprecedented speed and scale.

3) Implement through Intelligent Automation—Todays implementation is agile, powered by artificial intelligence, digital technology and machine learning. Today's workforce consists of digital workers. A combination of all these elements is what we call intelligent automation. The operations at workplace is set to have both human and digital workers, becoming an intelligent workforce

completing the task/set of tasks with the power of intelligent automation, analytics and decision-making capabilities while increasing the productivity and quality and cutting down the cost. In the process of intelligent workforce operations, cognitive technologies play an important role, learning from the best.

User Experience integrated with AI is delivering variety of digital fintech services like automated investment advisory, mobile wallet, online banking etc. These AI powered fintech solutions provide advice at the transactional level. Smart wallets like wallet.ai assists consumers analyze, price, and consider every single thing they spend money on, at a granular level that no human assistant could match.

The new age smartphones connected with IoT healthcare devices are allowing doctors to monitor patient's health from a distant place. These devices are enabling doctors to trace activity logging, heart rhythm and many other in-depth physical examinations of patient. Enterprises with specialized knowledge and advanced skillset in areas like analytics and healthcare delivery are interacting and competing through the common data and resource aggregation platforms. Infrastructure providers are providing the routine healthcare facilities.

Technology has enabled user smartphones to buzz and direct consumers, to the shop that is close to their physical location, basis the search that was performed earlier—"Nike shoes you looked up the other day is available in your size and color with 15% discount, at a store 300 meters away from you." Using machine learning algorithms, the app traces the likes and preferences of users through their browsing patterns and allows them to make a wish list. The app then sends notification to the user's smartphone when the user is near a store. Once the user enters the store, beacons mounted in the store detect the user's presence and credits the user with walk-in points.

AR (Augmented Reality) provides visual simulation and is significantly augmenting learning. Teachers are intensifying lessons and textbook content, engaging students by transforming the object or place being studied—rather transforming the classroom itself. AR is powering to supplement education across the curriculum and actively engage learners in a way that is meaningful and aligns with multiple learning styles. Artificial Intelligence (AI) is seen in most accustomed intelligent digital assistants like Google Home, Siri and others for help in everyday life.

Governments are propelling the growth of startups by providing incentives in the form of grants, tax benefits, and reimbursements. Governments Entrepreneurs Programme offers commercialization fund and business growth grants to budding entrepreneurs. The ever-increasing incentives for new businesses and investors, such as free trade zones and tax deductions have accelerated the investment in technology projects and are making many nations more appealing. These initiatives are seen as a development tool that encourages job creation and the investment of new capital to promote innovation. The newer startup policies and acts are making it simple to register a new company, relaxing labor regulations, creating tax incentives for investors in fast-track startups and innovation. The government across nations are encouraging tech startups, incubators, accelerators, venture builders and VCs in the digital transformational journey.

Adopting technology and innovation as the core capability is helping leaders to continue dominating their industries. Large well-known companies around the world are adopting robotics into their business models. AI platforms are automating business processes, from capturing client information to serving as the consumer interaction interface. Cognitive computing, machine learning, augmented reality is incorporating self-learning systems which use data mining, natural language processing and pattern recognition to imitate how the human brain functions. These are creating automated IT systems that solve problems without requiring human assistance. IoTs are encompassing many emerging technologies like connected cars, wearables, drones and robotics. Increasingly, connected sensors are being applied to heavy machinery, supply chains, and factories leading to new digitization. IoT is allowing industrial players to derive useful insights, make decisions, and optimize their systems by augmenting with worker wearables and collaborative robotics.

The year 2020 with the global pandemic is unprecedented for most of us. Businesses are struggling to survive, thinking of recover, and making plans to sustain and grow. Everyone is wishing that they could have been better prepared. Now is the time, when the leaders need to take care of the present, focus on recovery and prepare for resiliency in the future.

We all have seen how the lockdown has decimated the traditional businesses and lives of each and every one of the seven billion global population. Many businesses have already made a decision that work from home will be the "NEW NORMAL." Businesses are looking to automate using AI based solutions and digital workers, rather than human workers. Bringing in AI driven innovative solution is a must.

Businesses who have embraced innovative technologies have done extremely well. Numerous examples of winners should give us the motivation and sense of aggression for action. We all wish, we could have been more prepared earlier. The good news is that each and every one of us in our respective businesses can still leverage numerous innovations and the new technology, to get our businesses back on track in a fairly short period. We need to prepare for the recovery and get the return on investment in less than six months.

AI driven intelligent automation solutions, available from Tholons (www.tholons.com), and the world's leading consulting and technology companies are being deployed by world's visionary businesses. You, as a leader have the fiduciary responsibility to your employees, customers, families and your business to act NOW!

We are living in a very dynamic and exciting world, where the challenges and opportunities are immense. Let's go out there and re-design the experience, embrace innovators and drive change for success of enterprises.

Appendix A

Tholons Global Innovation Index™ 2021

Tholons Global Innovation Index is the leading touchstone to benchmark the accelerated digital transformation of industries and services globally. The index evaluates ranks and provides location strategies to multinational corporations, countries, governments, multilateral agencies, analysts and investors. Digital is a critical element in transforming industries globally. Digital innovation in emerging technologies, such as cloud, AI, big data and analytics, which saw unprecedented demand during the pandemic, will sustain even in the coming decade.

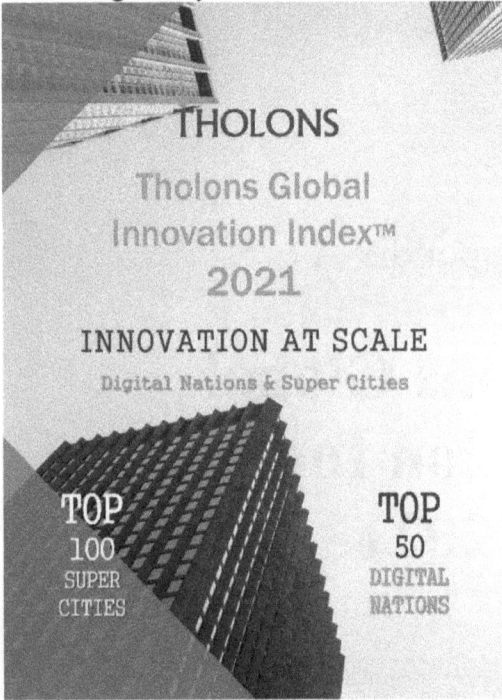

There is a need for enterprises to re-imagine new model for personalization, that emphasizes customer agency. Organizations will need more active engagement strategies, if they want to thrive and succeed. Leading businesses are adopting "human-AI" collaboration. As social distancing becomes the norm, in many industries, robots are transitioning faster than expected from regulated environments to unregulated environments. Corporations and governments are looking for more and newer "contact-less" solutions.

AI driven intelligent automation solutions, available from Tholons (www.tholons.com), and the world's leading consulting and technology companies are being deployed by world's visionary businesses. You, as a leader have the fiduciary responsibility to your employees, customers, families and your business to act NOW! We are living in a very dynamic and exciting world, where the challenges and opportunities are immense. Let's go out there and re-design the experience, embrace innovators and drive change for success of enterprises.

Tholons introduced innovation, startup ecosystem and digital transformation as key components to define its index for **Top 50 Digital Nations** and **Top 100 Super Cities** in 2017. The current index for 2021 has a much higher emphasis on digital innovation, re-imagining of consumer experience, transcendental knowledge, future readiness, digital competitiveness and diversity & inclusion.

The Business of Consumer Experiences

The pandemic has reshaped industries and the world. Retailers are reshaping their businesses in real-time, to survive, re-align and grow. Companies are enabling employees to work from home / remotely by providing assets, technology enabled collaboration and productivity tools. Data-driven recruitment by use of AI, is making traditional hiring process seamless through intelligent automation, from employee interviewing to onboarding. Let's look at how consumer experiences are being collated and transformed by disruptive innovation:

Shopping
Today's shoppers opt for an array of options; shopping in a store, shopping online, getting product reviews, friend's opinion and opting for product comparisons. New and improved technology seem to be a step behind always. Companies need to think much ahead to retain customers. They can do this by harnessing big data wherever available—be it through social media, surveys or just a plain old feedback. According to Statista, in 2019, retail e-commerce sales worldwide amounted to 3.53 trillion US dollars which is projected to grow to 6.54 trillion US dollars by 2022. One of the most common online activities worldwide is online shopping. Having developed into a full-fledged online store with a variety of products, Amazon is a highly data-driven analytics adopter. In order to come up with new, better goods and services, the company collects crucial data and analyzes them deeply.

Learning

Educational institutions sticking with traditional methodologies have been experiencing a decline in revenues. These institutions must implement digital learning that exceeds a set of new expectations. The demand for knowledge continues to be outpacing supply. In 1995, only 4% of the American Corporations used online learning vs 77% now. In order to educate the students, CX analytics should be used to understand student's perspectives. Leading companies are making the most of the latest digital tools to propagate, share and acquire knowledge. Their innovative methods include cloud-based gamification, interactive videos, social and mobile learning.

Staying Healthy

Everything to do with health - food, exercise, diagnostics, hospitals is to stay healthy and enjoy quality life. In today's world, the therapists, yoga teachers, doctors are all consulting online through video calls. People are conscious of what they buy - the medicine prescribed by doctors are being searched on internet to get more information before consuming. Online search attracts 3 times more visitors to hospital sites. Searching is by far the preferred way to locate healthcare providers online for patients. AI algorithms, like any medical practitioner, are able to design treatment plans and mine medical records faster than any existing player in the healthcare ecosystem. Integrated information from wearable tracking systems based on applications allows for personalized medical monitoring and treatment. A Harvard Business Review study found that, relative to their historically trained counterparts, VR-trained surgeons had a 230% increase in their overall results. Bio issues, artificial organs, pills, blood vessels can now be printed in 3D printing and the list goes on. Data, AI and VR based tools are giving the doctors and surgeons 10 times the power and knowledge than what they had earlier.

Entertainment

Customers pay for the value and experience that they want and get in the entertainment industry. Brands need to know exactly where their target audience spends their time and why. For both music and video, streaming has become the main revenue engine, transforming how the industry and artists distribute media to audiences. AI, AR, VR, blockchain, high internet speed and other such technologies and solutions have further helped improve content delivery, leading to an explosion in the growth of streaming services. Media companies are now able to analyze customer and behavioral data and recommend content in real time. My Disney Experience enables customers to thoroughly plan their vacations online; make dinner reservations, accommodations, explore the park using an interactive map and filter activities. Adaptive bitrate streaming technology is used by Netflix, streaming films and TV shows over the Internet to change the video and audio quality to fit the broadband connection speed and real-time network conditions of a consumer.

Travelling

Consumers are drifting to online marketplaces to thoroughly review and compare their options. In today's world, all it takes is a minute for travellers to tell hundreds of people how exactly their trip went, either on social media or on online review sites. Online feedback is the digital version of word of mouth and can make/break a company. Mobile phone has become our travel agency, best restaurant locator, tour guide, map navigator, and more. According to TripAdvisor, 45% of users use their smartphone for everything having to do with their vacations. It's possible to "teleport" ourselves to the most distant corners of the globe without getting off the couch with AR/VR having joined the travel universe.

Hiring

Organizations are now looking forward to the 'New Normal' being adopted. It is bound to place pressure on talent management teams and recruiting agencies as the market for talent gathers traction again. Many companies have achieved greater productivity and performance with remote work, and many workers appreciate the flexibility of time and do not have to fly to work on a regular basis. By using AI, data-driven recruitment renders the conventional recruiting process seamless through full automation, from employee verification to onboarding. Technology is powering payrolls, performance, and video interviews. 2021 will see an increase in the implementation of paperless and contactless onboarding technologies, including employee authentication. Since work from home is the new normal, businesses can shift towards project-based or freelance recruiting, and employers are free to recruit the best talent worldwide. In addition, this would also offer domain experts opportunities to take up projects on a short-term basis.

Paying

The digital mastery of players already dealing with change has been checked by the pandemic. Non-cash transactions are on a rapid growth path, accelerated during COVID-19 by enhanced adoption. In the midst of unparalleled growth, regulators seek to instil confidence and resolve the risk of non-cash payments as players cooperate to quench uncertainty. Emerging technologies are seen as an elixir for fraud prevention, data-driven offers are seen for value-added proposals, and distributed ledger technology focuses on digital currency solutions, enhancement of performance, and cost gains. New players are incorporating payments into their value chains, such as retailers/merchants, while technology companies are upgrading their financial services game as a central stage by weaving deals around payments. Constrained by budgets, business models such as Platform-as-a-Service (PaaS) are considered by businesses to provide cost-effective and superior customer experience. Payment market leaders work with companies outside their industry to obtain the opportunity to enhance the consumer experience. American Express customers can use their loyalty points to pay for taxi rides in New York.

Businesses who have embraced innovative technologies have done extremely well. Numerous examples of winners should give us the motivation and sense of aggression for action. We all wish, we would have been more prepared earlier. The good news is that each and every one of us in our respective businesses can still leverage numerous innovations and the new technology, to get our businesses on track in a fairly short period. We need to prepare for growth acceleration. Here is a list of few solutions that are being deployed by clients across the globe to manage and grow through the global crisis (www.tholons.com):

- Digital Omni-channel Contact Centers with Conversational AI
- Cloud Migration, DevOps, SecOps & WFH
- AI Driven Finance and Banking Solutions and Intelligent Automation
- AI Driven Healthcare Intelligent Automation Solutions
- Intelligent Supply Chain
- Automated SAP Migration and Operation
- Intelligent Online Product Posting and HR-Employee Onboarding
- Monetizing Intelligent Process Automation

Innovation at Scale for Enterprises

Innovation at scale for enterprises is a digital transformation process/function which brings three key elements to deliver sustainable impact:

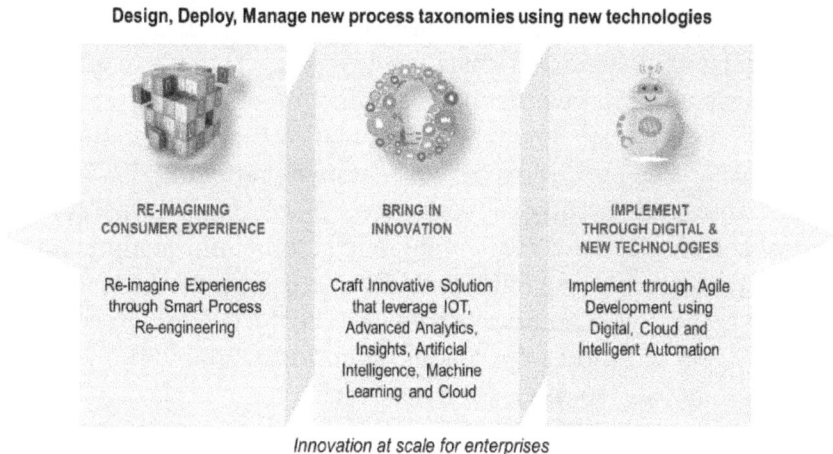

Design, Deploy, Manage new process taxonomies using new technologies

RE-IMAGINING CONSUMER EXPERIENCE	BRING IN INNOVATION	IMPLEMENT THROUGH DIGITAL & NEW TECHNOLOGIES
Re-imagine Experiences through Smart Process Re-engineering	Craft Innovative Solution that leverage IOT, Advanced Analytics, Insights, Artificial Intelligence, Machine Learning and Cloud	Implement through Agile Development using Digital, Cloud and Intelligent Automation

Innovation at scale for enterprises

- Re-image Consumer Experience—Enterprises are working with creative design studios to re-imagine consumer experience through a process of "dreaming." The idea is to take clients and consumers to a state of levitation that can enable them to dream the impossible. It is important to ensure that the dreamers are free from the shackles of technology, practicality and implementation realities. This creates a sweet but near impossible dream. How do we make the impossible, possible!

- Bring in Innovation—The world is full of innovators, who have re-imagined, re-designed and implemented amazing innovative solutions to deliver snippets of near impossible dreams. It is possible to explore, discover and bring in these innovations to make the impossible dreams, near possible. The solution design through innovation in most cases delivers the wow consumer/client experience. The innovative solution includes data and intelligent automation strategy, cloud, mobility, blockchain, AI, social media, IoT, machine learning and cybersecurity with advanced analytics to unlock new intelligence. The solutioning augments the power of humans with artificial intelligence, leverage AI and advanced analytics algorithms to sense, comprehend, act and learn across value chain at an unprecedented speed and scale.

- Implement through Intelligent Automation—Todays implementation is agile, powered by artificial intelligence, digital technology and machine learning. Today's workforce consists of digital workers. A combination of all these elements is what we call intelligent automation. The operations at workplace is set to have both human and digital workers, becoming an intelligent workforce completing the task/set of tasks with the power of intelligent automation, analytics and decision-making capabilities while increasing the productivity and quality and cutting down the cost. In the process of intelligent workforce operations, cognitive technologies play an important role, learning from the best.

Tholons Global Innovation COUNTRY INDEX – 2021
TOP 50 DIGITAL NATIONS

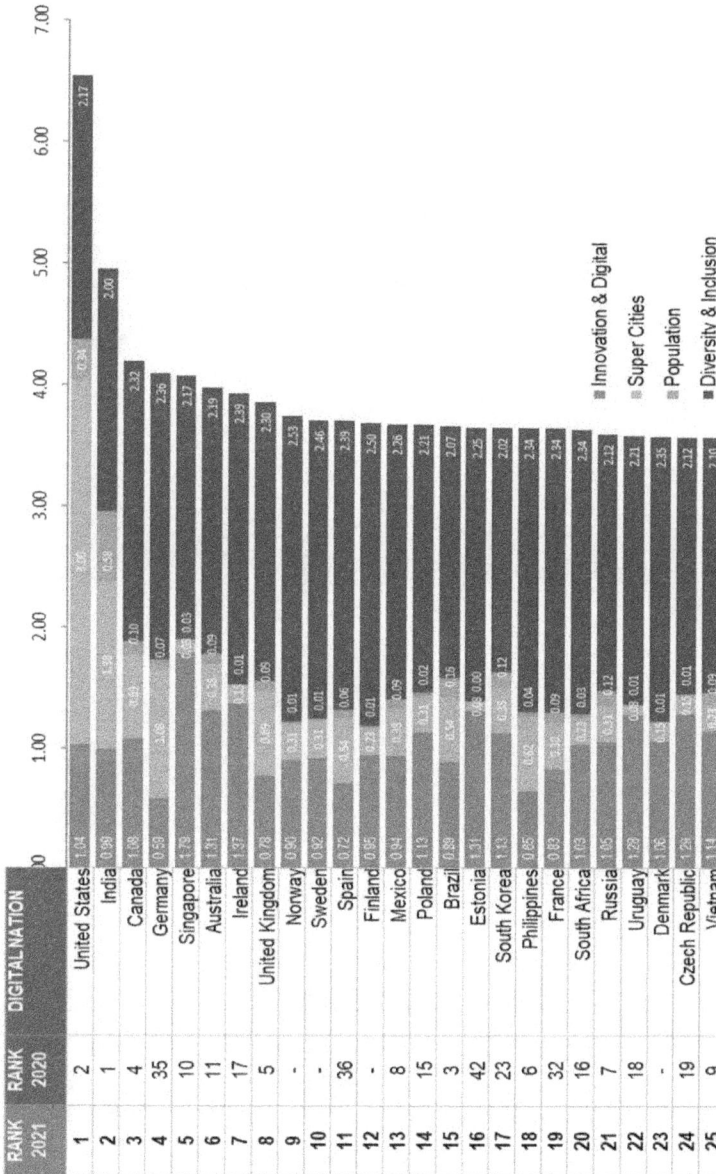

Tholons Global Innovation COUNTRY INDEX – 2021
TOP 50 DIGITAL NATIONS

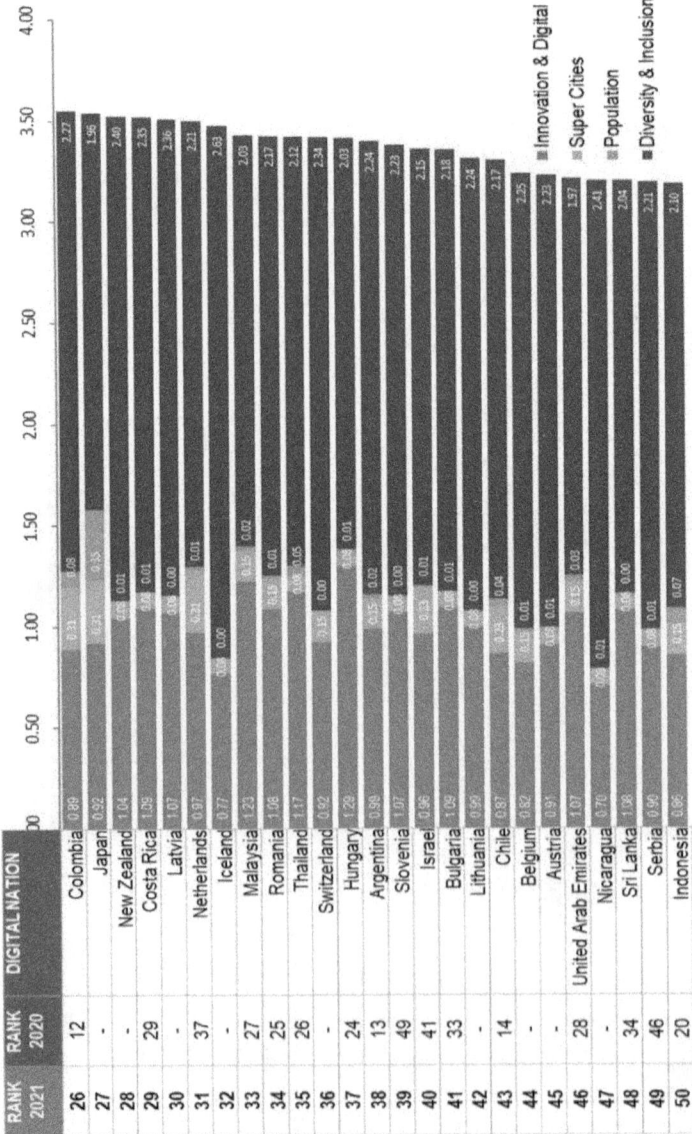

Tholons Global Innovation COUNTRY INDEX – 2021
TOP 100 SUPER CITIES

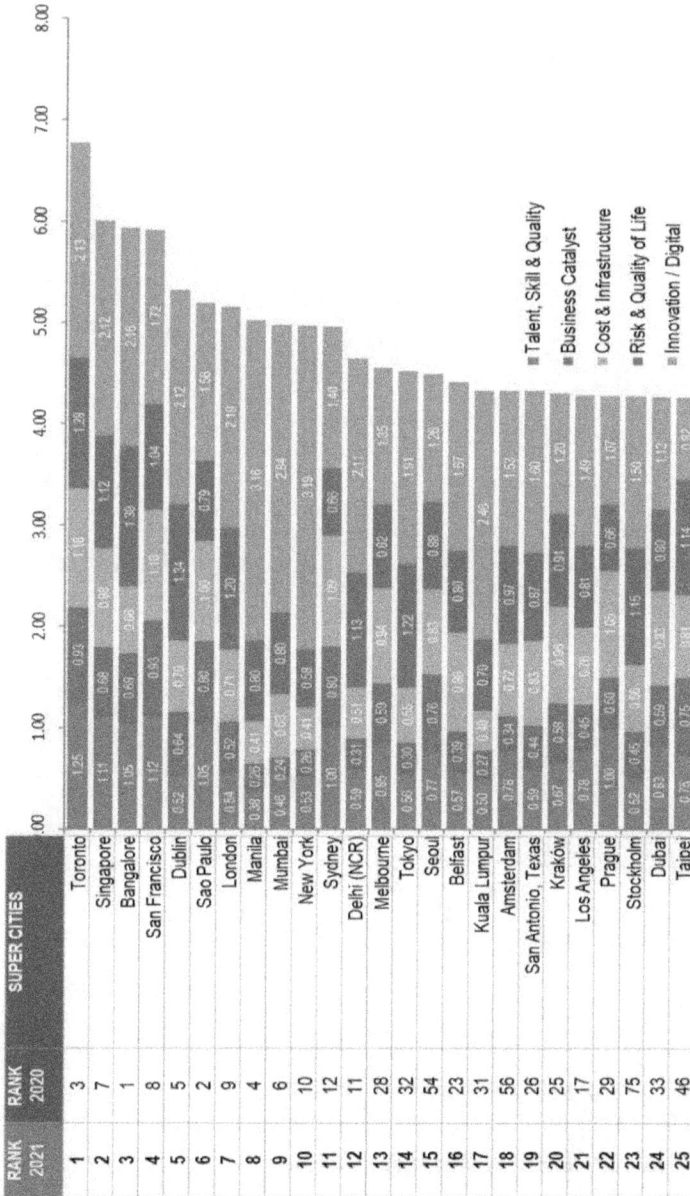

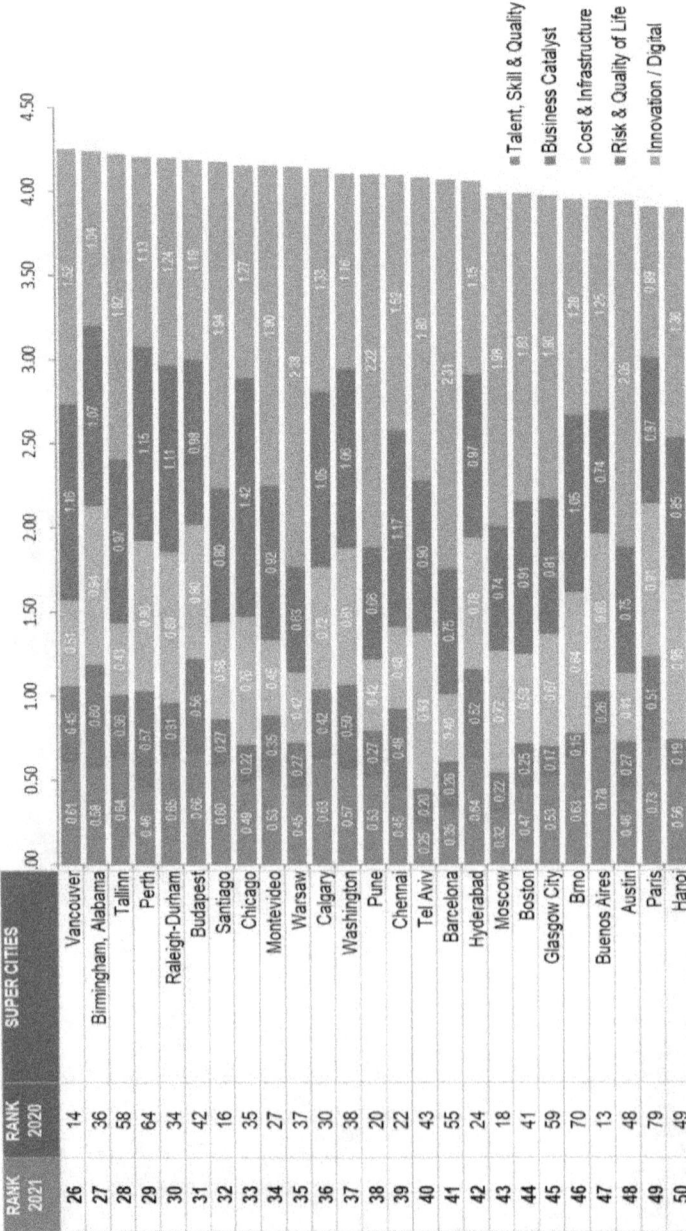

Tholons Global Innovation COUNTRY INDEX – 2021

TOP 100 SUPER CITIES

Legend: Talent, Skill & Quality · Business Catalyst · Cost & Infrastructure · Risk & Quality of Life · Innovation / Digital

RANK 2021	RANK 2020	SUPER CITIES	Talent, Skill & Quality	Business Catalyst	Cost & Infrastructure	Risk & Quality of Life	Innovation / Digital
26	14	Vancouver	0.61	0.45	1.16	0.51	1.52
27	36	Birmingham, Alabama	0.59	0.60	1.07	0.94	1.04
28	58	Tallinn	0.64	0.36	0.43	0.97	1.82
29	64	Perth	0.46	0.57	1.15	0.90	1.13
30	34	Raleigh-Durham	0.65	0.31	1.11	0.69	1.24
31	42	Budapest	0.66	0.56	0.98	0.90	1.18
32	16	Santiago	0.60	0.27	1.94	0.80	
33	35	Chicago	0.49	0.22	1.42	0.76	
34	27	Montevideo	0.53	0.35	1.90	0.46	
35	37	Warsaw	0.45	0.42	2.28	0.63	
36	30	Calgary	0.63	0.42	1.05	0.75	1.33
37	38	Washington	0.57	0.50	1.06	0.81	1.16
38	20	Pune	0.53	0.27	2.22	0.66	
39	22	Chennai	0.45	0.48	1.17	0.66	
40	43	Tel Aviv	0.25	0.20	0.90	0.49	1.80
41	55	Barcelona	0.35	0.26	2.31	0.75	
42	24	Hyderabad	0.64	0.52	0.97	0.78	
43	18	Moscow	0.32	0.22	0.74	0.72	
44	41	Boston	0.47	0.25	0.91	0.53	1.93
45	59	Glasgow City	0.53	0.17	0.81	0.87	1.80
46	70	Brno	0.63	0.45	1.05	0.64	1.28
47	13	Buenos Aires	0.78	0.26	0.74	0.63	1.25
48	48	Austin	0.46	0.27	2.05	0.44	
49	79	Paris	0.73	0.51	0.97	0.81	0.89
50	49	Hanoi	0.56	0.19	0.85	0.95	1.36

Tholons Global Innovation COUNTRY INDEX – 2021

TOP 100 SUPER CITIES

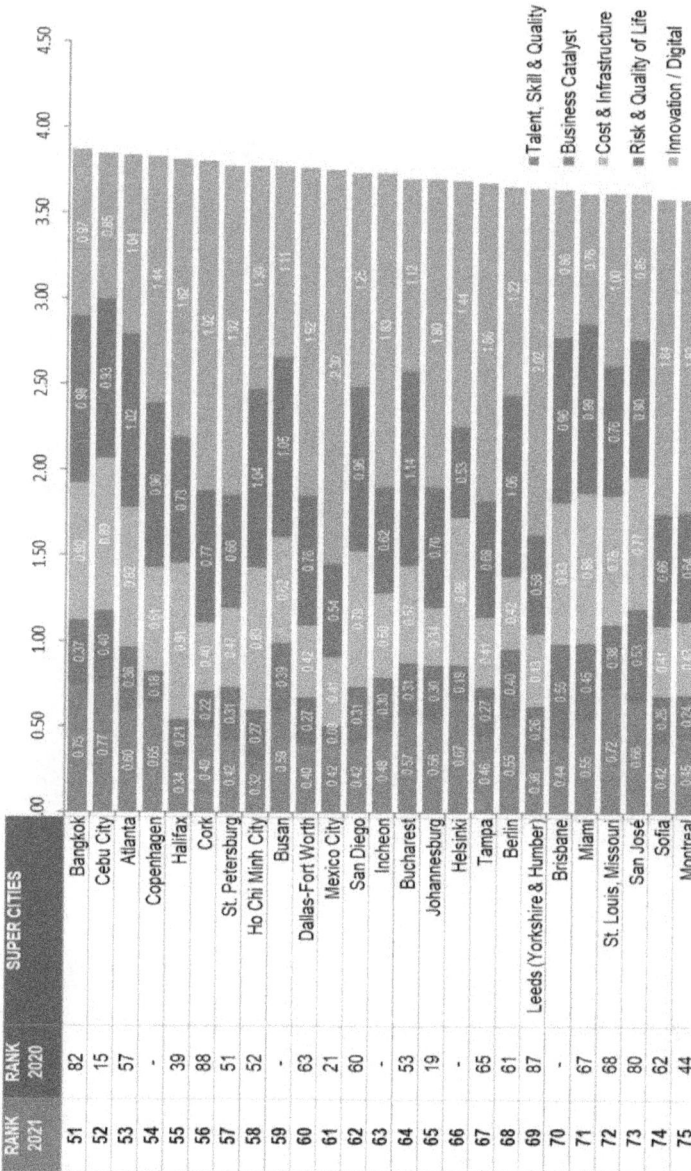

Tholons Global Innovation COUNTRY INDEX – 2021
TOP 100 SUPER CITIES

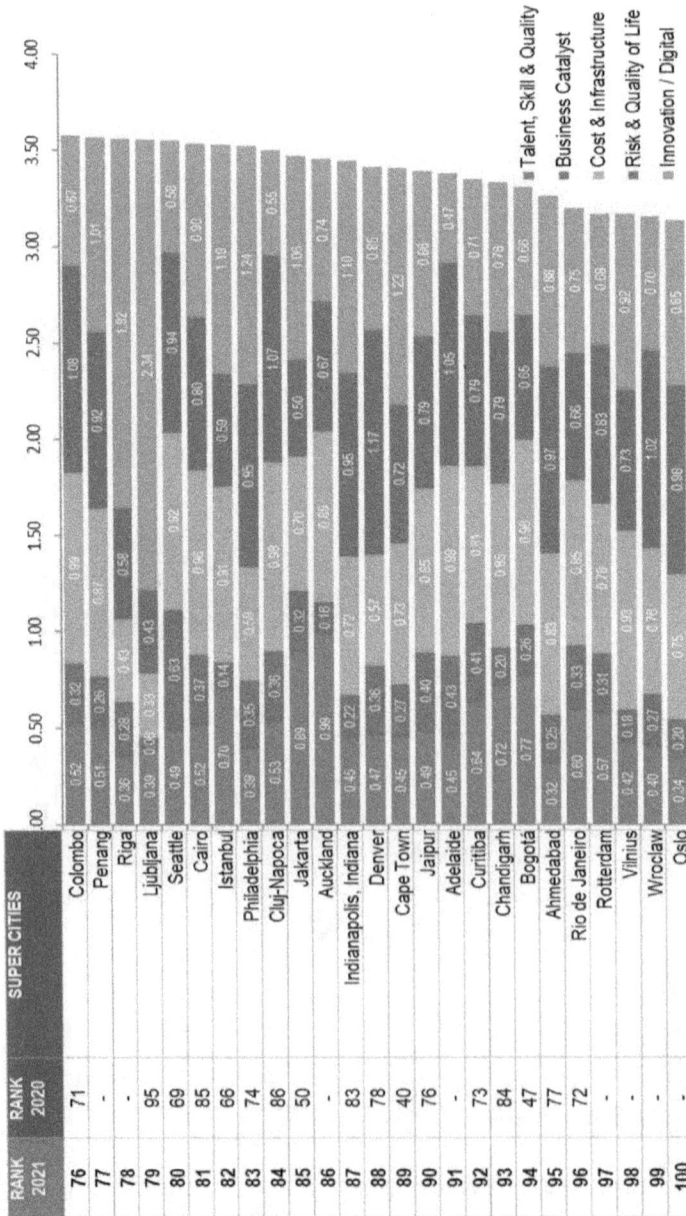

Legend:
- Talent, Skill & Quality
- Business Catalyst
- Cost & Infrastructure
- Risk & Quality of Life
- Innovation / Digital

RANK 2021	RANK 2020	SUPER CITIES	Talent, Skill & Quality	Business Catalyst	Cost & Infrastructure	Risk & Quality of Life	Innovation / Digital
76	71	Colombo	0.52	0.32	0.99	1.08	0.67
77	-	Penang	0.51	0.26	0.87	0.92	1.01
78	-	Riga	0.36	0.28	0.58	1.92	0.43
79	95	Ljubljana	0.39	0.38	0.33	2.34	0.58
80	69	Seattle	0.49	0.63	0.62	0.94	0.92
81	85	Cairo	0.62	0.37	0.46	0.80	1.18
82	66	Istanbul	0.70	0.14	0.31	0.59	1.24
83	74	Philadelphia	0.39	0.35	0.60	0.95	
84	86	Cluj-Napoca	0.53	0.36	0.98	1.07	0.55
85	50	Jakarta	0.69	0.32	0.70	0.60	1.05
86	-	Auckland	0.96	0.18	0.60	0.67	0.74
87	83	Indianapolis, Indiana	0.45	0.22	0.77	0.95	1.10
88	78	Denver	0.47	0.36	0.57	1.17	0.85
89	40	Cape Town	0.45	0.27	0.73	0.72	1.23
90	76	Jaipur	0.49	0.40	0.85	0.79	0.66
91	-	Adelaide	0.45	0.43	0.80	1.05	0.47
92	73	Curitiba	0.64	0.41	0.71	0.79	0.71
93	84	Chandigarh	0.72	0.20	0.85	0.79	0.78
94	47	Bogotá	0.77	0.26	0.96	0.65	0.66
95	77	Ahmedabad	0.32	0.25	0.83	0.97	0.86
96	72	Rio de Janeiro	0.60	0.33	0.65	0.66	0.75
97	-	Rotterdam	0.57	0.31	0.76	0.83	0.88
98	-	Vilnius	0.42	0.18	0.92	0.73	0.92
99	-	Wroclaw	0.40	0.27	0.78	1.02	0.70
100	-	Oslo	0.34	0.20	0.75	0.98	0.85

Tholons Research Methodology

Methodology

Tholons Global Innovation Index is the leading touchstone to benchmark the accelerated digital transformation of industries and services globally. The index evaluates, ranks and provides location strategies to multinational corporations, countries, governments, multi-lateral agencies, analysts and investors. Digital is now a critical element in disrupting and transforming industries globally. Technology, Business process management companies and multinational corporations need to align with the stark reality of digital innovation and transformation.

Tholons location assessment methodology integrates both primary and secondary research.

- **Primary Research:** Tholons utilizes surveys and interviews with service providers and buyers. These surveys are used to determine delivery and consumption trends for globalization services in specific destinations. Primary data gathering interviews were used to determine market and labor sizes as well as expansion strategies of leading service providers. Tholons utilizes its extensive network of industry stakeholders including buyers and suppliers of services, governments, trade bodies and associations to collect and validate data and analysis.
- **Secondary Research:** Tholons utilizes secondary research methodologies to gather volumes of historical data and various statistics and economic related data from governments, global institutions and agencies, and monetary bodies.
- **Quantitative and Qualitative Analysis:** Tholons employs a combination of quantitative and qualitative analysis in developing the weighted rankings. Our proprietary ranking framework continues to evolve to align with most of the current market realities and demand. Further, qualitative analysis was implemented to provide perspective to the quantitative results of the report. Tholons carefully considered numerous variables when providing final rankings, validated by senior thought leaders from Tholons and industry leaders globally.

Following are the metrics used to evaluate location attractiveness. The relative weights of each metric are based on their importance to the location decision, again derived from client experience and industry surveys. TGII's 2017 ranking was published with traditional factors having 80% weightage and digital with 20% weightage. TGII's 2018, 2019 and 2020 ranking attributed traditional with 75% weightage and digital with 25% weightage. The current TGII 2021 Report attributes traditional with 60% weightage and digital with 40% weightage. In addition, the report also covers "Diversity and Inclusion" measuring women's equality - pay gap, women in leadership role and funding made available to them as part of Digital Nations ranking.

TGII Super Cities Attributes and Weight

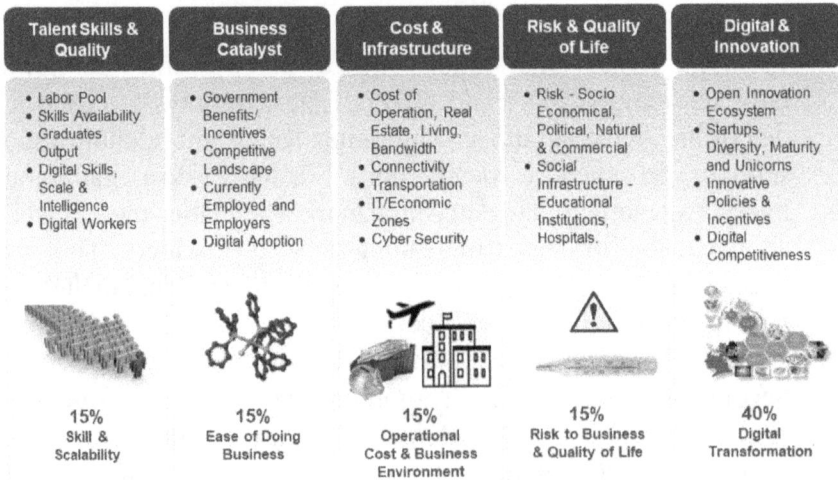

Talent Skills & Quality	Business Catalyst	Cost & Infrastructure	Risk & Quality of Life	Digital & Innovation
• Labor Pool • Skills Availability • Graduates Output • Digital Skills, Scale & Intelligence • Digital Workers	• Government Benefits/ Incentives • Competitive Landscape • Currently Employed and Employers • Digital Adoption	• Cost of Operation, Real Estate, Living, Bandwidth • Connectivity • Transportation • IT/Economic Zones • Cyber Security	• Risk - Socio Economical, Political, Natural & Commercial • Social Infrastructure - Educational Institutions, Hospitals.	• Open Innovation Ecosystem • Startups, Diversity, Maturity and Unicorns • Innovative Policies & Incentives • Digital Competitiveness
15% Skill & Scalability	15% Ease of Doing Business	15% Operational Cost & Business Environment	15% Risk to Business & Quality of Life	40% Digital Transformation

Talent, Skills and Quality

Refers to the overall talent pool availability in a particular location and in relation to the capability of the labor pool to meet staffing demands and fulfil outsourced services from both scale and quality perspectives. Total population, demography, labor pool size, annual tertiary graduate output, both in technical and non-technical background and skills proficiencies were among the host of related components considered when generating the Talent, Skills and Quality Score.

Scoring: *10 = High scale and quality 1 = Low scale and quality*

Business Catalyst

Business Catalyst measures industry-related activity as well as the degree of organizational support present in a location, which is geared to develop the services outsourcing industry and support the development of digital technology, innovation and entrepreneurship. Current industry performance including the top service providers and captives, location's headcount and revenues, etc. Number of Employees in ITO & BPO, Number of ITO & BPO Providers and ease of doing business and Policies and incentives for FDI are among the critical sub-components.

Scoring: *10 = Strong business catalyst 1 = Weak business catalyst*

Cost & Infrastructure

Cost includes relevant payroll and non-payroll costs in a location. Real Estate cost, basic outsourcing industry salary costs are among the factors considered to generate the Cost Score. With the internet now being indispensable, the bandwidth cost has also been included. Infrastructure refers to the availability of physical and technological platforms/systems, which are required to host outsourcing operations from a location. This considers the availability of office space, number of special economic zones or IT parks, mass transports systems, number of leased line providers, Number of Hospitals, Number of Educational Institutes and airport Connectivity.

Scoring: *10 = Low cost and Capable infrastructure available 1 = High cost & Inadequate infrastructure*

Risk & Quality of Life

Risk encapsulates the qualitative perceptions and measurable occurrences of natural and socio-political risks which in turn affect the quality of Life, where quality of life refers to non-operational considerations and ecosystem components that may affect living and working conditions. Risk is quantified according to the propensity of events to impact personal safety and the overall stability of a location to hosting business operations. The overall Risk and Quality of Life Score is generated from the identification of social infrastructure, non-work culture, and availability of leisure and recreational facilities, among others and from the identification of factors such as economic, political, natural, social and also cyber and digital risks in a location.

Scoring: *10 = Low risk& High quality of life 1 = High risk & low quality of life*

Digital & Innovation
Digital and Innovation is the lifeline of today's businesses. Businesses must embrace digital technologies and platforms like social media, mobile technology, cognitive computing, cloud and artificial intelligence to cater to the changing demands.
Following are the parameters considered for Digital and Innovation:Open innovation ecosystem, number of startups, startup diversity and maturity, innovative policies and incentives, unicorns, cybersecurity, global digital competitiveness, digital literacy rate i.e. the number of people using internet, digital evolution, digital talent and high tech patent grants, business agility, usage of RPA/AI/cloud, investors.
Scoring: *10 = High digital and innovation 1 = Low digital and innovation*

Super Cities in Top 100
Super Cities in Top 100 refers to the number of cities from a given country that are ranked in TGII Super Cities extensive list.
Scoring: *Number of cities in TGII Super Cities Index.*
10 = High Super Cities 1 = Low Super Cities

Workforce/Total Population
Workforce is a measure of the size of the workforce population in a given country and is an indicator of a talent pool available that can be skilled/re-skilled to serve cross industries in services.
Scoring: The total population of the country that is derived from the population of the cities present in top 100 list for a given country.
10 = High workforce 1 = Low workforce

Diversity and Inclusion
Diversity and Inclusion measures the women's equality, pay gap, women in leadership role and funding made available to women entrepreneurs.
Scoring: The score derives from metrics for pay gaps, women in leadership roles and funding for women entrepreneurs along with data and ranking from World Economic Forum, Global Gender Gap Index – 2020.
10 = High Diversity & Inclusion 1 = Low Diversity & Inclusion

Editorial Board

Debjani Ghosh
President
NASSCOM

"It has been a challenge for the entire industry. Employee safety, client business and well-being are utmost important. As India strives to carry on 'business as usual' while battling the devastating global outbreak as well as its economic fallout, the IT sector is leading the way with solutions for remote innovation and collaboration. Easing of compliance norms for 'work from home' for IT sector, innovation in the times of self-isolation, fears of large-scale job losses in the impending recession and India Inc.'s efforts to support the war on Covid-19."

Mihir Shukla
CEO and Co-founder
Automation Anywhere Inc.
USA

"Innovation at Scale is a must-read. Captivating. If you are going to lead in transforming and modernizing your business, this thoughtful book takes you on a journey of enterprise innovation – Cloud, Artificial Intelligence, Intelligent Automation and much more."

Kiran Mazumdar Shaw
Chairman & MD
Biocon Ltd.
India

"This Book "Innovation at Scale" is a must read for leadership teams who are driving business transformation within organizations."

Kevin Campbell
CEO
Syniti Inc.
USA

"Innovation at scale is a must read for CXOs & Clients of Global Corporations to understand and implement the new technologies across their enterprises."

Michael Barrett
Professor
IT & Innovation
University of Cambridge
UK

"Every leader must engage with visionary businesses to deploy an innovative AI driven intelligent automation solutions. It is the fiduciary responsibility to your employees, customers, families and your business to act NOW."

Andrew Wrobel
Chief Strategist
Emerging Europe

"Both manufacturing and services industry must continue to make smarter decision by using predictive data analytics, machine learning and augmented reality to survive in the arena of digitization fever."

Ankita Vashistha
Director
Investments & Innovation
Tholons Inc.
UK

"Artificial intelligence, innovation, digital marketing and tech-based entrepreneurship will fuel the economies in this decade. Untapped power of women entrepreneurship, engagement and empowerment is leading the way like never before."

ABOUT THE AUTHORS

Ankita Vashistha is the CEO of Tholons Capital, MyStepUp Entrepreneurship and Innovation Foundation and Managing Partner at StrongHer Capital, with over 12 years of experience in private equity, consulting, venture capital and innovation across UK, US and Asia. She founded and launched India's first venture capital fund, Saha Fund, to promote and invest in women entrepreneurship and technology. She is an engineering graduate and masters from Cranfield school of management, UK. Ankita works very actively in the startup ecosystem globally across US, UK, India, Singapore and Japan, to source, evaluate, mentor and invest in tech companies. She also works with portfolio companies to help them create value, scale and expand, leveraging technology.

Avinash (Avi) Vashistha is Chairman and CEO of Tholons and Managing Director at MyStepUp Entrepreneurship and Innovation Foundation and StrongHer Capital. He is the former Chairman and CEO of Accenture (India) and ran a multi-billion global delivery with a workforce of over 150,000. Avi is a highly accomplished, results oriented Chief Executive (CEO) and Venture Capitalist (VC) with 30+ years of experience in large global projects, CEO/board level strategy consulting, venture capital, digital transformation, and innovation.

Book Project Leader and Supporting Writer:
Anthony Rajesh, Project Head, Tholons Inc.

Appendix B

References

[1] Marine Energy, *"ORPC gets preliminary permit for Maine tidal project"*, 2016

[2] Juniper Research, *"Digital ad spend to reach $520 billion by 2023"*, 2019

[3] Tim Cooper and Mathew Robinson, *"Accenture Remaking customer markets unlocking growth with digital"*, Accenture 2015

[4] ICEF Monitor, *"Study projects dramatic growth for global higher education through 2040"*, 2018

[5] Simon Kemp, *"Digital 2019: Global Internet Use Accelerates"*, We are Social 2019

[6] Accenture, *"Accenture fact sheet q4 fy19"*, 2019

[7] Source: Innosight, Creative Destruction Whips through Corporate America, http://www.innosight.com/innovation-resources/strategy-innovation/creative-destruction-whips-through-corporate-america.cfm

[8] Farnell element14, *"Smart Factories: Handling complexity with flexibility"*, 2017

[9] Kerry A Dolan, *"Cement meets the cyberworld"*, Forbes 2017

[10] Dave Evans, *"The Internet of Things How the Next Evolution of the Internet Is Changing Everything"*, Cisco 2011

[11] Forrester, 01/2016: https://go.forrester.com/blogs/hadoop-is-datas-darling-for-a-reason/

[12] The English Oxford Living Dictionary, 2019 Edition: https://www.lexico.com/en/definition/artificial_intelligence

[13] AIIM: https://www.aiim.org/What-is-Robotic-Process-Automation#

[14] McKinsey & Co, 03/2017: https://www.mckinsey.com/business-functions/mckinsey-digital/our-insights/intelligent-process-automation-the-engine-at-the-core-of-the-next-generation-operating-model

[15] Horses for Sources, July 2019: https://www.horsesforsources.com/rpa_dissatisfaction_get_real_080119

[16] Forbes, 2018: https://www.forbes.com/sites/londonschoolofeconomics/2018/09/25/why-we-should-learn-to-love-the-robot/

[17] Technology Review, 2019:
https://www.technologyreview.com/s/612582/data-that-illuminates-the-ai-boom/

[18] AI Index, 2018:
http://cdn.aiindex.org/2018/AI%20Index%202018%20Annual%20Report.pdf

[19] Medium, 2019: https://medium.com/dataseries/artificial-intelligence-and-recent-billion-dollar-investments-2019-759e78b042ad

[20] TechCrunch, 2019: https://techcrunch.com/2019/06/24/gartner-finds-rpa-is-fastest-growing-market-in-enterprise-software/

[21] Gartner, 2019: https://www.gartner.com/en/newsroom/press-releases/2019-06-24-gartner-says-worldwide-robotic-process-automation-sof

[22] SAP, 2018: https://news.sap.com/2018/11/sap-acquires-contextor-robotic-process-automation/

[23] Forbes, 2019:
https://www.forbes.com/sites/tomtaulli/2019/11/15/microsoft-aims-to-upend-the-rpa-robotic-process-automaton-market/

[24] International Federation of Robotics, *"IFR World Robotics Report"*, *2018*

[25] Grand View Research, *"BPO Market Analysis Report and Segment Forecast 2018-2025"*, 2018

[26] Kevin McCaney, *"Marines' trail-blazing unmanned helicopter returns home"*, Defense systems 2014

[27] Blizx Robotics, *"Robotics Trends: 5 trends to watch"*, 2015

[28] James Carroll, *"Robotics and automation market in Germany reaches new heights"*, Vision Systems Design 2019

[29] MarketsandMarkets, *"Artificial Intelligence Market by Offering Technology, End-User Industry, and Geography - Global Forecast to 2025"*, 2018

[30] Accenture, *"Boost Your AIQ Transforming into an AI Business"*, 2017

[31] https://news.sap.com/2017/12/idc-study-s4hana-seven-insights/

[32] Statista, *"Internet of Things - number of connected devices worldwide 2015-2025"*, 2016

[33] Fortune Business Insights, *"Internet of Things (IoT) Market Size, Share and Industry Analysis Forecast 2019 – 2026"*, 2019

[34] Jacob Morgan, *"A Simple Explanation Of 'The Internet Of Things'"*, Forbes 2014

[35] MarketsandMarkets, *"Smart Home Market by Product Software & Services, and Region - Global Forecast to 2024"*, 2019

[36] Dan T Pickett, *"The Internet of Everything Will Impact Everything, Including Your Next Tech Job"*, Wired 2015

[37] Andreas Mai, *"Smart Connected Vehicles: Driving to the Bottom Line!"*, Cisco 2015

[38] Lisa Jerram, *"How Will Wireless Connectivity, Vehicle Autonomy, and Electrification Converge with On-Demand Mobility?"*, Navigant Research 2016

[39] Jay Samit, *"4 Technology Trends That Will Transform Our World in 2018"*, Fortune 2017

[40] LexisNexis Risk Solutions *"H2 2018 Cybercrime Report"*, ThreatMetrix 2018

[41] Panda Security, *"27% of all recorded malware appeared in 2015"*, 2016

[42] Steve Morgan, *"CV-HG-2019-Official-Annual-Cybercrime-Report"*, Cybersecurity Ventures 2018

[43] Evan Perez, *"US takes out computer malware that stole millions"*, CNN Business 2014

[44] Simon Kemp, *"Digital 2019: Global Internet Use Accelerates"*, We are Social 2019

[45] Ericsson, *"5G Consumer potential"*, 2019

[46] Lakshna Rathod, *"Cost of a Data Breach: Ponemon Institute Report"*, Diligent 2019

[47] Verizon, *"2019 Data Breach Investigations Report"*, 2019

[48] Phil Blackman, *"Global Fintech investment hits record $111.8B in 2018"*, KPMG 2018

[49] Mordor Intelligence, *"AI in Fintech market – growth, trends, and forecast (2019-2024)"*, 2018

[50] Laura Shin, *"How Millennials' Money Habits Could Shake Up the Financial Services Industry"*, Forbes 2015

[51] EY, *"Global Fintech Adoption Index 2019"*

[52] Taavet Hinrikus, *"Executive Summary: The Future of Finance"*, TransferWise 2016

[53] William Alden, *"BBVA Buys Banking Start-Up Simple for $117 Million"*, The New York Times 2014

[54] J. Clement, *"PayPal: active registered user accounts 2010-2019"*, Statista 2019

[55] Invoicex, *"Unbundling of finance"*, 2015

[56] Ilya Pozin, *"15 Fintech Startups to Watch in 2015"*, Forbes 2014

[57] Crunchbase: *"Currencycloud"*

[58] Oliver Smith *"Currencycloud's Secret Pipes Are Powering Europe's Billion-Dollar FX Businesses"*, Forbes 2018

[59] Tom Davies, *"The Unbundling of Banks – Are Businesses Services Next?"*, SME Finance Forum 2016

[60] Consultancy.uk, *"Integral and agile strategy is key for success of Fintech Investments"*, 2017

[61] KPMG, *"Fintech 100 2018 report"*

[62] Stefan Biesdorf and Florian Niedermann, *"Healthcare's digital future"*, McKinsey 2014

[63] HealthCare Global, *"Top 10 Healthcare Startups delivering real value"*, 2014

[64] Business Wire, *"Healthcare Global Market Opportunities and Strategies to 2022"*, 2019

[65] Mordor Intelligence Report," *Telemedicine market- Growth, Trends, and Forecast (2019-2024)"*, 2019

[66] The Journal of mHealth, *"Global Digital Health 100"*, 2018

[67] Sharon Goldman, *"5 cutting-edge retail technology trends"*, CIO 2015

[68] Grand View Research, *"Connected Retail Market Analysis and Segment Forecasts to 2022"*, 2016

[69] Business Insider Intelligence, *"Amazon accounts for 43% of US online retail sales"*, 2017

[70] BusinessWire, Global Higher Education Testing and Assessment Market 2018-2022, *Adoption of Digital Badges to Boost Demand*, Technavio

[71] Technavio, *Top 10 Edtech Companies in the World*, 2018

[72] BloombergNEF for the United Nations Environment Program and Frankfurt School's UNEP Center, 2019

[73] Energy Digital, *Top Trends in Energy Sector*

[74] Capgemini, *"The Digital Talent Gap Developing Skills for Today's Digital Organizations"*, 2017

[75] NCFA *"Canada is North America's up-and-coming startup center"*,2017

[76] www.startupindia.gov.in/

[77] fundsforNGOs, "Ireland's Best Young Entrepreneurs Programme inviting Innovative Business Idea & New Start-up", an article, 2019

[78] Darrell Etherington, Japan's government is providing nearly $1B to boost homegrown space startups, TechCrunch, 2018

[79] Alice TruongMay, *"These are the types of startups that have been hammered by "down rounds""*, Quartz 2016

[80] Ann Williams, *"Singapore overtakes Silicon Valley as No. 1 for global start-up talent"*, The straits times 2017

[81] Lauren Davidson, *"How Sweden became the startup capital of Europe"*, The Telegraph 2015

[82] SIG Fiduciaire, *"Incentive programs-small businesses in Switzerland"*, Sigtax

[83] Ben Adams, *"J&J opens the doors to its new biotech hub in Houston"*, Fierce Biotech 2016

[84] *"Start-up Companies"*, The University of Texas MD Anderson Cancer Center

[85] Elizabeth MacBride, *"Dubai Has $300 Million To Entice The World's Best Startups To Its Accelerator"*, Forbes 2016

[86] Dan Mangan, *"SoftBank and Saudis launch largest tech investment fund ever, announce $93 billion in committed capital"*, CNBC 2017.